The Infanticide Controversy

The Infanticide Controversy

Primatology and the Art of Field Science

AMANDA REES

The University of Chicago Press
Chicago and London

Amanda Rees is lecturer in sociology at the University of York.

The University of Chicago Press, Chicago 60637
The University of Chicago Press, Ltd., London

Printed in the United States of America

18 17 16 15 14 13 12 11 10 09   1 2 3 4 5

ISBN-13: 978-0-226-70711-2 (cloth)
ISBN-10: 0-226-70711-3 (cloth)

Library of Congress Cataloging-in-Publication Data

Rees, Amanda.
The infanticide controversy : primatology and the art of field science / Amanda Rees.
p. cm.
Includes bibliographical references and index.
ISBN-13: 978-0-226-70711-2 (cloth : alk. paper)
ISBN-10: 0-226-70711-3 (cloth : alk. paper) 1. Infanticide in animals.
2. Primates—Behavior. I. Title.
QL762.5.R44 2009
599.8′15—dc22
2008054955

⊗ The paper used in this publication meets the minimum requirements of the American National Standard for Information Sciences—Permanence of Paper for Printed Library Materials, ANSI Z39.48-1992.

To my grandmother, Brenda Williams

CONTENTS

ACKNOWLEDGMENTS

In the first instance, my thanks must go to the primatologists who gave up their time and energy to talk to me about the infanticide debates. I thank them for their unfailing good humor, and also for their patience, when I insisted on repeatedly revisiting what to them seemed blindingly obvious. This project really would not have been possible without their cooperation. Second, I would like to express my gratitude to the British Academy, which provided essential financial support for this research, and the University of York (UK), which funded a year-long Anniversary Lectureship in which the final draft of this book was completed. Third, I thank my colleagues, some of whom helped by reading drafts of chapters, others of whom helped by forcing me to explain (not always successfully) why they should bother reading drafts of chapters. In particular, I thank the two anonymous reviewers of the entire manuscript, whose comments and queries have, in my opinion, improved the book enormously, and I thank John Forrester, who supervised the initial research on which this book is based. Fourth, I am grateful to the editorial team at the University of Chicago Press for everything they have done in preparing this book for publication. Finally, I would like to thank my husband, Iwan Rhys Morus, who arrived on the scene some time after this book had been begun, but who rapidly became quite necessary to its successful completion. Diolch yn fawr, fy nghariad.

INTRODUCTION

The Infanticide Controversy

I've only seen one, and it happened very quickly, and I wasn't expecting it. I did not go to the field expecting to study infanticide. . . . When it did happen, what a bizarre mix of emotions the incident evoked. I'm thinking, wow, I just saw a group member kill an infant, or maul it, and I wasn't entirely sure what had happened. It was clearly a scuffle on the ground. And then there's the infant horribly wounded and dying and three animals are running away. In my papers, I've always tried to be careful to clearly note that I'm not a hundred percent sure who did it. My gut feeling has always been that it was the dominant female. But I have to be perfectly honest that it happened so quickly that all I knew was that these three other individuals were in the immediate area running away from where the scuffle had occurred. And here I am, the objective scientist, saying, "Oh, my god." I knew not to interfere. One of the adults stayed with the infant for about an hour, trying to pick up the infant, and staying with it, giving alarm calls. So I went some distance off. While all this is happening, my brain's saying "This is cool," but I'm also breaking into tears at the same time. This was the little infant I had been watching for almost a month. I felt such a mix of emotions, that this is a life, and it was not quite a personality yet, but an individual that I knew, and yet also knowing that I had seen something really amazing, that nobody expected to see.

—Leslie Digby, *Duke University*

It's really a horrible thing, and what's . . . even more horrible, is when you see the male come in and they just sort of stalk down the mother. . . . That's the most frightening bit because you don't know and you can see this male with these babies in his sights and you know what's about to happen, and it makes you feel terrible, it really does make your stomach churn, literally.

—Louise Barrett, *Liverpool University*

The slaughter of the innocents is always horrifying. Although I have never been to the field and have never seen wild primates with my own eyes, I have watched videos and films of chimpanzee and lion males killing infants and have been appalled and disgusted by what appears, to my unfortunately anthropomorphic gaze, to be the cold-blooded butchering of helpless babies. Perhaps I am particularly squeamish; perhaps living in urban settings has insulated me from the harsh, practical realities of unmediated nature: in any case, it is clear that I am not alone in reacting with nausea and distaste to these violent events. But infanticide, or infant killing,[1] is also a fascinating intellectual phenomenon. It is a rare event, but one that has now been reported across a very wide range of species;[2] yet the subject remains controversial in one particular area of behavioral biology and for one particular order of animals—the Primates. The controversy turns on whether infanticide is a pathological response to particular social and environmental conditions or represents an adaptive strategy, forming part of the standard animal behavioral repertoire. Given the horrific nature of the event itself, it is not surprising that it has been the subject of controversy: what is surprising is that this controversy, which has now persisted for almost four decades, has been largely confined to the primatological literature. Why was it that primatology, alone among the disciplines of behavioral biology, found the idea of adaptive infanticide so contentious?[3]

One reading of the debate could explain its persistence by the peculiar characteristics of primatology. Primatology stands out among the observational sciences of animal behavior for two reasons. First, its heritage was multidisciplinary, with practitioners receiving their early training in fields as diverse as mathematics and sociology. Great effort was needed, particularly in the early years of the discipline's emergence, to avoid confrontations between researchers who were operating with very different expectations regarding, for example, standards of proof or the relation between theory and method. Second, and perhaps more significant, the subjects of the primatological gaze were the primates, humanity's closest relatives. The intense public and media interest in the discipline's discoveries meant that primatologists also found themselves constantly confronting laypeople on what, for less charismatic megafauna, might be their uncontested intellectual terrain. In addition, the discoveries primatologists made about primate behavior were inevitably treated as directly relevant to the understanding of human behavior. Unlike other behavioral scientists, from the outset primatologists found themselves and their research in the public eye and subject to political appropriation.

Those primatologists who were directly involved in the controversy also cited a range of personal reasons for the debate's persistence—reasons rooted in what sociologists Nigel Gilbert and Michael Mulkay (1984) have called the "contingent repertoire." Scientists on both sides of the debate explained the controversy by reference to a combination of emotional and political responses. Those who treated the event as an adaptive strategy were being "terribly macho," while those who considered it pathological or accidental either were committing the moralistic fallacy (assuming that what should not happen could not happen) or were simply unwilling to believe that animals could act in such a brutal and repulsive way purely to further selfish interests. Accusations of anthropomorphism, historically the cardinal sin of behavioral biology, were flung from both sides: supporters of the adaptive strategy were "clearly" imputing to the animals both intent and conscious calculation, while opponents were "obviously" motivated by the morphological and genetic similarities between human and nonhuman primates. After all, if it is "natural" for our closest relatives to behave this way, surely it should be so for humans as well? Those unconvinced by the adaptive hypothesis argued that this was yet another example of gratuitous sociobiological storytelling, and one with potential eugenic overtones: those for whom the hypothesis made perfect sense ascribed the existence of opposition to the feminist movement, to the "science wars," and to the postmodern turn in the social sciences. Personal hostilities, selfish career-advancing decisions, and the deliberate misreading of the opposition's position were attributed to those on all sides of the debate. But there is nothing new about these accusations. They echo those made in the wider debates within academe regarding the application of sociobiology to human behavior, and perhaps more important, they have their parallels in the scientific controversies that have been examined by sociologists, historians, and anthropologists since science studies began.

But what happened in the debates surrounding the act of infanticide or infant killing was about much more than unpleasant behavior by a few peculiar primates, be they monkeys, apes, or humans. Quite the contrary: the controversy centered on how one should go about doing science in the field; how one should conceive of the relation between theory and data in a discipline oriented in the first instance toward descriptive analysis; how Darwinian evolution was to be defined and practiced; and whether humanity was to be treated as a part of the natural world or apart from the rest of it. At the very heart of the debate were the issues of observation and authority: How much of a particular interaction needed to have been seen, and how often, and in how many places and in what circumstances before it could be considered a representative behavior of a given species? In this sense, then, the infanticide debates both reflected and focused key theoretical themes and methodological questions, issues that had been present in primatological practice since the inception of modern field studies of apes and monkeys in the mid-twentieth century. And yet—infanticide still stands out as a special case. Apparently similar disputes occur frequently in the observational sciences, particularly in the field of animal behavior, but they do not usually lead to the intractable altercations that characterized the infanticide debate. Indeed, what surprises an outside observer is how quickly such questions normally are either resolved or temporarily shelved. And the infanticide case is especially intriguing when one considers that most, if not all, of the criticisms initially raised against the evidence for adaptive infanticide could have been raised against most, if not all, of the fieldwork being conducted at that time.

This paradox is precisely what makes the infanticide debate particularly interesting for the history of science. The topic may have become controversial because of the contingent issues (the emotional, the political) that surrounded it—but it remained controversial because of the particular conditions under which observational, behavioral science is conducted in the field. As such, it provides a wonderful opening for exploring in detail the way field scientists go about doing science in the field. As later sections of this introduction will demonstrate, most of the work carried out in the history and sociology of science has focused on laboratory science. While there are similarities in the way researchers approach field and lab space, there are also important discontinuities, particularly with reference to the relation between observation, theory, and interpretation and the way field sites must be not only physically and socially created by their users but also maintained over a considerable time in the teeth of geopolitical and environmental conditions that can be extremely chaotic. With a very conscious debt to the work of Harry Collins (1992, 2007; Collins and Evans 2002), this book is ultimately an exploration of what will be called the "fieldworker's regress," as it is expressed through the peculiarity of place.

Caveat

The rest of this introduction will outline the infanticide controversy as it developed from the mid-1970s to the early years of the twenty-first century and will consider the theoretical and methodological background to this study as well as summarizing the book's structure and its key arguments. However, I will begin with a caveat. One of the more unfortunate results of the science wars of the late twentieth century was the presumption that those who studied science intended to undermine it.[4] For most of us, nothing could be further from the truth. Science and scientists represent one of the most potent and authoritative forces in the world today, making it critical that the processes through which scientific knowledge is produced be more widely understood. Controversies provide useful windows on the ways scientific knowledge is created. They allow us to watch and listen as scientists contend over competing interpretations of some aspect of the natural world. Students of controversy, by and large, have two choices. Either they can investigate controversies in the past that have been resolved in favor of one side or the other, or they can examine subjects still under debate.[5] Clearly I have taken the latter path, but this has its dangers. Most pressing, in investigating an ongoing controversy one risks that the investigation itself will be taken to be an intervention in the course of the debate.[6] This is emphatically not what this book is intended to do. I am a historian and a sociologist, not a primatologist or a biologist, and I have no business making any judgments as to the truth or falsity of any claims about the origin or maintenance of animal behavior. Instead, my interest lies in the strategies and procedures by which claims are made about the truth or falsity of data or theory. As the very existence of scientific controversy demonstrates, facts do not speak for themselves but are spoken for by scientists—otherwise controversies, disagreements, and debates between scientists could not occur. I do not by any means dispute the existence of fact: I am interested in the period between the making of a claim about the natural world and its acceptance as factual by the relevant scientific community. During this time its "factual" status is unclear, and what one scientist considers indisputable evidence for a given statement can seem to another to be a fundamental misapprehension of the problem. To reiterate: I am interested in investigating the competing judgments of the scientists involved and their grounds for reaching these conclusions. Nothing here is meant as a personal opinion on the abilities or accuracy of those who participated in the infanticide controversy. I cannot say who is right or wrong; I can only observe the debates with fascination and respect.

Studying Scientific Controversies

Since this matter is so fundamental to understanding the approach I have taken, a short review of the emergence of this approach may be appropriate. The study of scientific controversies has been a basic strategy of the history and sociology of science since Thomas Kuhn's *Structure of Scientific Revolutions* (1962). Using examples drawn from the history of astronomy, chemistry, and physics, Kuhn demonstrated that the sciences did not advance through the incremental accretion of knowledge but were marked by periodic dramatic transformations in what could be known about the world. During "paradigm shifts" such as those associated with the work of Copernicus, Newton, and Lavoisier, what was to be counted as "factual" depended on which worldview one held. Each of these shifts was a "successful" revolution, in that the new worldview gained sway over the old, even as numbers of natural philosophers were persistently unwilling to accept the new perspective. For Kuhn this was not a matter of stubborn prejudice. He knew that extraordinary claims demand extraordinary evidence: rejecting the assumptions that have ordered one's professional life is not to be taken lightly. And crucially, it was not that the new worldview was "true" and the old "false." Instead, both interpretations were trying to give as complete an account as possible of why the world worked as it did, and different researchers, depending on their particular positions and interests, would rate these attempts as more or less successful or plausible.[7]

The "strong program" in the sociology of science[8] that emerged in Edinburgh in the 1970s strongly emphasized this last element of Kuhn's approach to the study of science. David Bloor's *Knowledge and Social Imagery* (1991/1976) made the principle that was to become known as methodological relativism central to the sociological investigation of science, insisting that judgments on scientific truth or falsity were no part of the sociologist's remit. In investigating the workings of scientific research, he demanded that researchers treat their subject matter impartially and symmetrically. In practice this meant rejecting the idea that true beliefs (because they are true) need no explanation, whereas false beliefs must be explained by sources external to scientific debate, such as personality, politics, or economics (were it not for these factors, false belief would not exist). Instead, Bloor insisted, similar explanations had to be sought for both true and false beliefs. As another science studies scholar put it, "Since the settlement of a controversy is *the cause* of Nature's representation not the consequence, *we can never use the outcome—Nature—to explain how and why a controversy has been settled*" (Latour 1987, 99; italics in original). Incidentally, this was another reason the study of ongoing controversies became popular. It is hard to remain impartial and insist on symmetrical explanations when one "knows" which side of a debate has been shown to be right.

While other traditions in the sociology of science, such as the ethnographic studies of science in action inspired by Bruno Latour and Steve Woolgar's *Laboratory Life,* concentrated on the day-to-day practices of scientists, engineers, and doctors in their working environments, the empirical relativist school associated with the work of Harry Collins concentrated on the theory and practice of present-day scientific controversies.[9] Three main insights into the nature of scientific research arose from the work of these researchers: the notion that much of the practice of science could be characterized as tacit knowledge;[10] the identification of a "core set" of scientific practitioners in relation to the potential resolution of controversy; and the notion of the "experimenter's regress." This last, put as simply as possible, refers to the demonstration that, in practice, it is never possible to replicate an experiment precisely. Therefore, even though the concept of replication plays a fundamental role in traditional philosophy of science, when it comes to scientific conduct it is always possible to show that the "replication" is not identical with the original.[11] Although in most cases this is not a problem, it becomes extremely important when there is controversy, for apparently trivial differences between experiments can then be seized on to demonstrate the inaccuracy of an opponent's interpretation of events. In this way, Collins's analysis of the experimenter's regress demonstrated that experimental results, rather than representing a touchstone for judging the truth or falsity of a theoretical claim, would readily yield contradictory interpretations where controversy participants held different worldviews. Hence the importance of having a core set of researchers capable of deciding when a controversy had run long enough and of ending it by various means. Through their investigations of laser building, parapsychology, and gravity waves, Collins and his coworkers were able to demonstrate the contingent aspects of scientific laboratory practice.

Clearly, Collins's analysis of the experimenter's regress was itself immensely controversial.[12] Traditional philosophy of science had long held that replication was central to the scientific method, one of the most important elements in demarcating science from other forms of human activity. Previous descriptions of scientific conduct had, broadly speaking, treated experiment and its replication as relatively straightforward processes whereby hypotheses could be tested against the natural world using scientific apparatus. Where problems did arise, these could be explained by flaws in the equipment, experimenter's error, or even inappropriate use of the laboratory space itself (Sibum 1995). But vitally, these were errors. At the heart of the traditional interpretation of science was a commitment to the idea that, in principle, the nature of the apparatus or the location of the laboratory should not matter. Doing the same things in the right order should produce the same results. The laboratory had, in fact, been created as a space where this could be done, a controlled space where universal standards could be applied regardless of location,[13] and scientists working in the laboratory could expect this to be true. The demonstrations by Collins and others that this was not so—at least where the subject under study was novel or where unexpected or extraordinary claims were being made (Collins 2004)—were themselves unexpected and unwelcome. But if this was the situation in the laboratory, then what of the field site? Did the same standards and expectations apply?

Studying Science in the Field

There have been relatively few attempts to study examples of controversial field science.[14] It can sometimes be difficult to draw a line between the field and the laboratory, especially where validating scientific knowledge moves aspects of that knowledge from the lab to the field and back again (Latour 1988), but conducting science in the field is nonetheless very different from doing so in the laboratory. Henrika Kuklick and Robert Kohler's introduction to their edited volume of studies of field science expresses this concisely when they point out that "unlike laboratories, natural sites can never be exclusively scientific domains" (Kuklick and Kohler 1996, 4). Workers at field sites and field stations must constantly manage a number of threats to the physical and epistemological integrity of both their sites and their persons, threats rarely experienced inside the four walls of the laboratory. This is particularly true when, as with primatology, field sites are in areas of the world where issues of conservation and development may conflict with research aims. Their ability to exert control over the laboratory notwithstanding, scientists are seldom the most important or the most powerful people at their field research sites.

When the science in question is ethological in orientation, again as with primatology, navigating the situation can become even more complex. At the heart of both ethology and primatology is a commitment to observing the behavior of an animal in its natural environment (Burkhardt 2005; Haraway 1990; Rees 2001a). This commitment can cause scientific controversy for one central reason: no consensus exists on the place of humanity within the "natural" world. As a general example of this, reported observations from a specific field site are often presented to their audiences as representative of the typical behavior of this species, but the acceptance of these observations as authoritative and typical hangs on observers' success in steering past two key questions.[15] First, have the animals under observation been affected either by human proximity or by the impact of humanity on the local environment? Second, is the reported behavior an account of the researcher's observations of the animals, or has the report been colored by assumptions about how this behavior might be interpreted? This second question, of course, becomes ever more important when the animals under investigation are the nonhuman primates, and it was fundamental to the development of the infanticide controversy. Whether a given field site could be treated as an example of pristine nature was essential to the unfolding of the debates, as was whether behavior observed in a single, possibly unique place could be treated as characteristic of the species. Methodological and theoretical questions that centered on attempts to transform the field site into a scientific space run through the history of the dispute, both as background and as a prime focus for controversy: Was animal behavior being altered by the methods researchers used to study them? Were researchers going to the field only to find what their theories led them to expect? And political and pragmatic realities placed constraints on what, if anything, researchers could do in particular places. Not only did research permission have to be regularly negotiated, but relationships with local administrators and people had to be carefully managed to shore up the (often physically nonexistent) site boundaries. In other words, unlike scientists working in the laboratory, field scientists constantly confronted the need for interpretive flexibility.

This issue of interpretive flexibility is at the heart of the notion of the fieldworker's regress. For obvious reasons, the standard of replication—so central to the ideal of the laboratory and the scientific method—does not and cannot apply to science in the field. Replicating field results is both literally and conceptually impossible: the functional alternative for the field is a combination of repetition and the comparative method. This might involve, for example, comparing and contrasting phenomena observed at different times, in different places, and under different conditions. However, the question of similarity and difference here—especially with regard to place—is obviously a matter of subjective interpretation filtered through the tacit knowledge gained through field experience, as the following chapters will show. Where a phenomenon or a behavior is novel or unexpected and may be questionable, then, as in the case of the experimenter's regress, it is normally accounted for by reference to the specific conditions under which it was observed. But significantly, field scientists also call on these conditions to explain why other observations are treated as *reliable* descriptions of the natural world. That is to say, the particular and distinctive characteristics of the place where phenomena are observed form an essential element in the reception given *both* to unexceptionable *and* to potentially controversial results—they are a central part of the normal practice of fieldwork. Field researchers are constantly called on both to reflexively evaluate their own observations and the conditions under which they were made, and to account for any differences between their work and that produced by other workers. This state—the fieldworker's regress—is the default state of the observer in the field and, ironically, accounts for why intractable controversies are relatively infrequent in field science, since all fieldworkers are aware that their own work is influenced by interpretation, limited visibility, uncertainty, and best estimates. The debates surrounding infanticide or infant killing in primates therefore represent an unusual situation where controversy has broken out, and this enables us to examine directly the problems—and benefits—of doing science in the field.

This controversy was first examined by Donna Haraway in 1983. Her account focused on how debates figuring the nonhuman primates had been materially shaped by contemporary political discourse. In this book I take a different tack, using infanticide as an example of a field-based scientific controversy and considering how the context of the field both aids and hampers the conduct of controversy. As I indicated earlier, I will use this controversy to examine the fieldworker's regress as it is expressed through the particularity of place.[16] Although at first the infanticide controversy may seem to be an unusual and unfortunate consequence of some of the peculiar characteristics of primatology, a closer examination of the debates reveals the far broader complexities inherent in scientific field practice.

Infanticide, Sociobiology, and Primatology

So what began the infanticide controversy? Aggressive attacks on infants that had led to either their death or their disappearance had a long history in langur field studies, as chapter 3 will show, and a range of explanations had been tentatively proposed to account for this odd behavior—but they were not initially controversial. Dissent and impassioned disagreement developed after three key events: Sarah Hrdy's use of sociobiological concepts in her interpretation of the behavior of the langur monkeys she watched at Abu, India; the publication of E. O. Wilson's synthesis of these concepts in 1975 and the responses to his attempt to apply them to human beings; and Hrdy's decision to seek out a wider audience for her adaptive explanation of primate infanticide. Clearly, sociobiology played a crucial role here, not just in relation to the furor it caused in academic circles, but also by the way it changed primatological practice in the field.[17]

For many primatologists, sociobiology was the fundamental factor that allowed the study of nonhuman primates to progress from descriptive natural history to analytic natural science. For example, Robin Dunbar described his encounter with sociobiology as "almost like going through a curtain round the mid-seventies. I did my two big field studies on primates in the early seventies, one was done in 71–72 and the other was done in 74–75 [and] they're on different planets, intellectually. You can just see the difference over that short period, . . . my whole thinking had been changed completely. Before it was rather descriptive, very traditional, and afterward it becomes very strategic, question driven if you like."[18] But primatologists' opinions on sociobiology's usefulness to primatology, given in retrospect, were not uniformly welcoming: for a small group of researchers, questions remained about the way sociobiological theories and concepts had influenced the development of field practice.[19] While no one doubted that the ideas of adaptation, kin selection, reproductive competition, and reciprocal altruism had had an immense impact on the direction primatology developed in the 1970s and beyond, this small group thought it had also introduced a more insidious danger. For them the move from what can broadly be characterized as "description" to "analysis" raised the specter of finding only what you sought: some researchers worried that, by going to the field with specific questions, primatologists might be blinded both to events themselves and to alternative explanations. Although there is no doubt that part of the critical response to the adaptive infanticide hypothesis was provoked by the controversy in which Wilson's *Sociobiology* was rapidly engulfed,[20] the broader questions that sociobiology raised for primatological field practice were just as vital.

This was not least because the hypothesis that infanticide was the product of sexual selection, of competition between males for access to reproductively active females, was one of the earliest examples of the sheer cerebral clout of sociobiological theories and concepts. It demonstrated that this rare, shocking, and violent behavior was not necessarily pathological, or even particularly unusual. There were circumstances when it might be expected: one could regard infanticide as an ordinary aspect of the behavioral repertoire in socially living animals.[21] Hrdy's 1974 paper was the first to make the explicit link between infant killing, male-male competition, and reproductive success. She proposed that infanticide would occur in situations where males encountered unfamiliar females with unweaned infants, enabling a male to maximize his own reproductive success at the expense of another: in other words, it was an adaptive, sexually selected strategy, paralleled by identifiable female counterstrategies that had coevolved to minimize the damage to their own reproductive quotient. In this initial paper, Hrdy was able to weave a range of observations of langur behavior into an account of langur society that had this act of adaptive infanticide at its heart: infanticide was neither random nor pathological but a predictable and logical event that would be evoked in particular social circumstances—a picture of langur life dramatically different from that described by her predecessors in the field. Three years later, after the storm of debate surrounding sociobiology was fully under way, Hrdy took her adaptive thesis to a wider audience of academics and interested laypeople (Hrdy 1977a, 1977b).[22] In these publications she emphasized her wider hypothesis—that infanticide was a reproductive strategy characteristic of the primates, which could be expected to occur regularly wherever certain conditions were met. This assertion provoked an instant response from Phyllis Dolhinow, who had carried out the first field studies of langurs in India and had described langur society in very different terms: for her, the behavior of Hrdy's langurs was unnatural and could not have been produced by natural selection (Dolhinow 1977; Curtin and Dolhinow 1978).

At issue during this stage of the controversy were two related problems: drastically dissimilar definitions both of primate social structure and of the kind of data needed to explore and test these theoretical constructs. Two models of langur social structure were being proposed. Hrdy's model strove to explain all observed langur behavior as examples of the same underlying pattern of male competition and female counterstrategy. Dolhinow's account insisted on a clear distinction between observations made at sites where the monkeys were behaving "naturally" and those where human influence had changed the conditions of life so radically that the animals were behaving "abnormally." Ultimately the question was whether the sites where infant attacks had been seen were places where representative langur social behavior could be observed. In addition, two perspectives on the appropriate practices of watching primate behavior in the field were visible. Specifically, the question was raised of how far it was appropriate to infer unseen behavior based on discontinuous observations. Only three infant deaths in total had actually been witnessed by scientists, all at a single site at Jodhpur.[23] Hrdy's model of adaptive infanticide depended on piecing together the (sometimes fragmented) accounts of what had actually been seen and showing that the patterns she identified at these separate sites both resembled each other and corresponded to the behaviors predicted by the sexual selection hypothesis. In contrast, Dolhinow and her graduate students laid great emphasis on how little was known of the conditions under which these infants were dying or disappearing: shaky grounds on which to erect a hypothesis about a primatewide behavioral strategy.

At the same time, however, as the following chapters will show, observing behavior in the field is a difficult and frustrating business. At this point, incomplete reports of behavioral observations were frequently being accepted as factual by the relevant communities, and events that were rarely observed, such as predation, were commonly taken to have exercised a powerful influence over the evolution of primate societies. It appeared that an impasse had been reached on the accuracy of the sexually selection hypothesis: insistence on detailed and contextualized observation had met the search for general patterns of behavior, and neither was prepared to compromise.

Consolidation of a New Paradigm?

The next year, 1979, saw the appearance of one of the most (in)famous descriptions of infant killing among the primates. Jane Goodall had been studying the chimpanzees of the Gombe Stream Reserve since 1960 and had found that their lives were far more complex than had ever been imagined. Her reports had emphasized their relatively gentle and friendly social interactions, occasionally interrupted by brief conflicts that, though intense, rarely resulted in serious wounding. In the mid-1970s, however, came a series of violent, eventually fatal attacks on members of the Kahama chimpanzee group by the neighboring Kasekela group. These events were disturbing in themselves, especially for researchers who had hoped to find in chimps solutions to the problem of uncontrolled human aggression (Goodall 1979, 598); but far worse in some ways was the behavior of two Kasekela females. Passion and Pom, mother and daughter, were repeatedly seen to kill and eat infants belonging to the other females of the community. From 1974 to 1976, only a single infant survived. The description of Passion and Pom's behavior, as it was recounted in *National Geographic,* horrified the magazine's international readership as much as it did those who had seen the events occur, but notably, Goodall refused to speculate on the reasons for their infanticidal acts (Goodall 1979, 613), stating only that far more research would be needed before conclusions could be drawn.[24]

Other researchers were less reticent as the debate continued over the nature and significance of primate infanticide. From 1974 until 1984, more than two dozen papers dealing with infanticide in one form or another were published in the primatological literature. Most either accepted the adaptive hypothesis or were favorably inclined toward it,[25] but during this same period it appeared that at least one of the "infanticidal" sites was no longer so: the team of German researchers now working at Jodhpur had failed to confirm the initial observations of infanticidal acts there (Vogel and Loch 1984).[26] This period of primatological debate ultimately culminated in a conference devoted to the study of infanticide in animals and man, sponsored by the Wenner-Gren Foundation[27] and held in August 1982 at Cornell University, New York (Hausfater and Hrdy 1984). This conference showed that the controversy over the evolutionary interpretation of infanticide and infant attacks appeared still to be confined to the primate literature—or more accurately, to the langur literature.[28] For other areas of behavioral biology, ethology, or zoology, the sexual selection hypothesis in particular and the notion of treating infanticide as adaptive in general had been eagerly welcomed.[29] But primatology still seemed bogged down in the questions raised in the early debates. Despite this mutual recalcitrance, Sarah Hrdy and Glen Hausfater, the editors of the collected conference proceedings, were determinedly upbeat. They were convinced that the adaptive thesis would shortly become conventional wisdom, a conviction that would come back to haunt Hrdy (Hrdy 2000, xi).

The decade following the conference saw a substantial array of papers describing events that their authors considered best explained by sexual selection or intrasexual competition.[30] By the early 1990s, although by far the majority of observed infanticides had been seen for Hanuman langurs,[31] reports of adult males' killing infants in circumstances thought to correspond with those predicted by the sexual selection hypothesis existed for Old World monkeys (langurs, colobus, blue monkeys, redtail monkeys, rhesus macaques, savanna and hamadryas baboons), New World monkeys (red and mantled howler monkeys, capuchins), and the great apes (gorillas). By this point a substantial body of literature describing the experimental documentation of infanticide by rodents in the laboratory had also built up, as well as a number of reports of infanticide in free-living social carnivores, in birds, and possibly also in cetaceans.[32] Although the reports of infanticide in langurs often still referred to the debates of the late 1970s, these references and citations were rarer in descriptions of infanticide in other primates and appeared even less frequently in relation to other taxa.

One might suggest that, as Hausfater and Hrdy had predicted in 1984, the infanticide controversy had been resolved resoundingly in favor of the adaptive hypothesis, as a case in which, for the first time, primatology had been ahead of the behavioral biology curve: a phenomenon first recognized in primates had now been acknowledged in many other species.[33] It is certainly true that by the early 1990s the presumption that infanticide had long roots in the primate evolutionary past had become sufficiently entrenched within primatology for infanticide risk to be pinpointed as a key factor in the evolution of primate social structure (Van Schaik and Dunbar

1990). From being a rare, pathological event, infanticide had apparently become the source primatologists must look to in order to explain one of the abiding problems of primatology: Why, unlike most other mammals, do primate males and females live in relatively permanent social groups?[34] However, events at another conference held in Erice, Sicily, in June 1990 demonstrated that the study of infanticide itself was still mired in doubt, suspicion, and debate (Parmigiani and Vom Saal 1994). Despite the apparent scientific consensus, participants feared that political and cultural reactions to the topic of infanticide were damaging scientists' ability to study it. Concern about this situation led some conference participants to prepare a statement on academic freedom, which was reproduced in the edited collection of papers along with facsimiles of the participants' signatures.

In addition, a new and vocal opponent of the adaptive hypothesis had appeared. As critics had done in the 1970s, Robert Sussman and his colleagues (Bartlett, Sussman, and Cheverud 1993) had examined the specific context of each report of infanticide and reduced the number of observed infanticidal attacks to only forty-eight cases across all primate species, nearly half of which had happened at a single study site—Jodhpur. Of these forty-eight, only eight met the hypothesis' predictions exactly. Was it realistic, they asked, to treat behaviors that appeared so rarely, and seemed largely confined to one particular place, as exemplars of an evolved, adaptive strategy?

A Few Peculiar Primates

What is notable about this paper, which attacked the adaptive interpretation of primate infanticide by returning to the particular context of the alleged cases of infanticide, is that it had its own particular context of publication. Whereas previous debates on the "adaptiveness" of infanticide had been conducted in journals devoted to the study of the behavior of primates and other animals, such as *Folia Primatologica, Primates, Behaviour,* or *Zeitschrift für Tierpsychologie,* as well as more general science outlets such as *American Scientist,* this one appeared in an anthropological journal. This shift is important in understanding the way this reopening of the infanticidal "black box" was interpreted by primatologists and other behavioral biologists at the time. Many were surprised and shocked that someone had felt it necessary to revisit this debate. For Charles Nunn, who was a graduate student when the paper appeared, "it was an eye-opener, that somebody was questioning it." For Robin Dunbar, the authors were "clutching at straws." Louise Barrett recalled that "it was kind of surprising to me that it just flared up again in this way . . . just when you think, oh, it's all sorted out." For these researchers, and others, as chapter 7 will show, this was an attempt to force a retrograde step, to return to arguments that had been shown to be out of date and out of data. For them, that the article had appeared in an anthropology journal said it all: although anthropology had been one of the most important parents of primatology,[35] and though many primatologists, especially in the United States, could be found in anthropology departments, the discipline, again especially in the United States, had never quite recovered from the fissures produced by the sociobiology controversy. In their opinion it was still reeling from the science wars and the influence of postmodern thought.[36] For supporters of the sexual selection hypothesis, anthropology's status as "science" was itself contested: no wonder then that a paper so out of tune with contemporary biological thought could appear in one of the discipline's key journals.[37]

However, Sussman and his colleagues had taken the critique of the adaptive hypothesis a step further, focusing explicitly on the biological basis of infanticide. They took issue with what they considered the "most fundamental assumption[s] of the sexual selection hypothesis" (1993, 981): that infant killing has a genetic, heritable basis; that it is differentially distributed (polymorphic) throughout a given population; and that those males that possess this trait leave more offspring than males that do not. From the point of view of this new generation of critics, supporters of the adaptive hypothesis had profoundly misunderstood what they needed to do to satisfy the skeptics. Bob Sussman, around whom the new wave of critique coalesced, described his intense frustration at this situation:

> I sit there and beat my head against the wall because basically every article says, well, now we've shown that this particular example of infanticide meets the immediate criteria for sexual selection. What they've missed is that these specific criteria have to be met to begin with. If they are not met, then you have no theory at all. In reality, the basic criteria needed to prove sexual selection are related to population genetics. . . . They miss that whole point, and we explained this in our article . . . after talking about the specifics . . . we must then ask how does a particular example relate to evolutionary questions and . . . how would you measure it evolutionarily?

From this perspective, supporters of the hypothesis have not even begun to gather the kind of information needed to test their model thoroughly and robustly.

Supporters of the hypothesis did not initially respond to this demand for further evidence. They were both surprised and shocked that what seemed to them such strong evidence in favor of the adaptive interpretation was once again being questioned, and the appearance of the article in an anthropology journal was telling. This appeared to be a marginal viewpoint associated with an area of intellectual life that had never fully embraced biological or evolutionary elements as explanatory factors. But two years later, in 1995, Sussman and his coauthors summarized their 1993 article in *Evolutionary Anthropology* (Sussman, Cheverud, and Bartlett 1995), a review journal that was primarily oriented toward biological rather than cultural anthropology. In this instance they were able to provoke a reply. This came, albeit with some initial reluctance, from Sarah Hrdy, Carel van Schaik, and Charles Janson.[38] They accepted that more data, both observational and experimental, would be useful. But for them the infanticide debate had now devolved into the opposition between "two different world views, both of them defendable" (Hrdy, Janson, and Van Schaik 1995, 154), one emphasizing specificity (of context and species), the other similarity (between patterns and across the comparative literature). An impasse, it seemed, had been reached yet again.

At this point it may be useful to note that the next year a conference in Teresopolis, Brazil, was devoted primarily to discussing how and why ideas about primate society had changed over the relatively short time during which primates had been systematically studied (Strum and Fedigan 2000). In the collected papers of that conference, the topic of infanticide is referenced only three times, a very small number for a phenomenon that had been identified both as the potential source for primatology's disciplinary cohesiveness and as a fundamental factor in the evolution of primate society.[39] For those primatologists who were writing the history of primatology, sexually selected infanticide remained a question rather than a solution in the study of primate society. As the twentieth century drew to a close, an attempt was also made in the pages of *American Anthropologist* (now edited by Bob Sussman)[40] to topple one of the essential pillars of supportive comparative data for the sexual selection hypothesis: Craig Packer and Anne Pusey's records of the lion population of the Serengeti (Dagg 1999). Anne Dagg, the author of this article, used much the same tools to unpick the lion literature as earlier critics had applied to the primatological literature and argued that the context of each case of infant killing did not fit the "type case" of infanticide as she interpreted it. This unexpected attack, unsurprisingly, provoked a furor (Silk and Stanford 1999; Shea 1999). In the same year Phyllis Dolhinow reentered the debate with an article (Dolhinow 1999) in which she tried to link the accounts of aggressive interactions between adult males and infants in langurs to infants' interest in aggressive interchanges between adults: on this reading, infants were not being targeted by males, they were simply getting in the way of very excited adults.

In the meantime, however, at a conference in Japan in 1998, Charles Janson and Carel van Schaik were discussing putting together a third edited collection of papers on adaptive infanticide; Janson because he was experiencing very hostile responses from reviewers to a paper he and a colleague had written that touched on infanticide, Van Schaik because of the opportunity a book would afford to review all the new infanticide data that had been identified over the past decade—data that included not just the observation of behavior, but also the physiological and genetic information that was becoming available through new field methods. In the preface to Van Schaik and Janson's book (2000), Hrdy rather ruefully looked back at the optimistic assertion that she and Glenn Hausfater had made in 1984: that within a decade, the adaptive nature of infanticide would be taken for granted. While this prediction, she felt, had come true for most of biology, for "those with backgrounds in the social sciences, perhaps especially in my own field of anthropology, it was wildly optimistic" (Hrdy 2000, xi).

Accounting for Difference

Indeed, as chapter 6 will show, many of the supporters of the sexual selection hypothesis pointed to this apparent rift between areas of intellectual life to explain why a minority of researchers continued to disagree publicly with the idea that infanticide could be explained by sexual selection. In this context, it is important to remember that primatology developed out of the intersection of the interests of disciplines ranging from anthropology to zoology, including psychology, medicine, mathematics, and sociology among others.[41] But the relative influence each of these fields had on the development of the study of nonhuman primate behavior depended to some extent on national traditions: so, for example, while the influence of anthropology was significant in the United States,[42] in Europe the stimulus for work on apes and monkeys grew out of a more ethological or zoological tradition (Hinde 2000). In a sense these specialties were initially not so different: both anthropology and ethology were originally observational disciplines, and they shared a fundamental commitment to understanding the social or cultural world from within that world: animals and humans should both be studied "naturalistically." But over the latter part of the twentieth century, some respondents argued that anthropology's unwillingness to embrace the new evolutionary synthesis—indeed, what was perceived as a general unwillingness to engage with any argument that related behavior to biology—meant that a profound misunderstanding of evolution, and what it meant to think in evolutionary terms, had developed within the discipline.

For some supporters of the adaptive hypothesis, this misunderstanding explained why the resurgence of overt controversy in the 1990s seemed largely confined to the United States, where the influence of anthropology over primatology had always been strongest and where the disputes over the implications and purposes of sociobiology had been most acerbic. Either the controversy originated in an inability to understand exactly what supporters of the hypothesis were arguing, because of the difficulty of transcending early training in different intellectual paradigms, or it was based on a deliberate misunderstanding of what the hypothesis predicted and when it might be applied, maintained out of obstinacy or an a priori commitment to particular political perspectives.[43] According to this reading, infanticide was singled out for attack not because the evidence for this behavior was significantly weaker than that for, say, predation on primates, but because infanticide was morally wrong, and to say that it was both natural and to be expected for nonhuman primates implied that it was an appropriate behavior for humans as well. To supporters of sexual selection, these researchers had fallen victim to the "naturalistic fallacy."[44] While researchers had different opinions on how valuable this debate had been,[45] not surprisingly those who supported the sexual selection hypotheses were unanimous in asserting that it was time to move on. In fact, one of the reasons a number of researchers mentioned for accepting the adaptive explanation for infanticide was that it had opened whole new areas for study and exploration. Taking sexually selected infanticide as a given meant they could embark on many more exciting projects.

In many ways these projects had been prefigured by Sarah Hrdy in her initial exploration of the consequences of taking infanticide seriously as a selective force, but as she and Glenn Hausfater had suggested in 1984, some researchers still saw "these speculations as the constructions of sand turrets on sand castles" (xxiii). Rather than moving the field forward, they were heading for a dead end. The opponents of the sexual selection hypothesis not only challenged the particular interpretations of how one should think about and study evolution provided by the sexual selection camp, but also questioned the significance of their minority status. Again not surprisingly, they denied that they had allowed their politics to color their science or that they had fallen into the cardinal behavioral sin of anthropomorphism. They proclaimed their commitment to Darwinian evolution as the means for understanding animal behavior, but they disputed the nature of the evidence that needed to be brought to bear on evolutionary hypotheses, particularly in understanding how behavioral traits might realistically be selected for in a given population over time, and how far particular events (infanticide) could be considered in isolation from wider behavioral complexes (aggression). As primatology in the United States escaped from the dominance of institutions where it was saturated with sociobiological influences, they argued, more young researchers would begin to question the unproven assumptions that underlay the application of this approach to wild, socially complex creatures like primates. Finally, people would begin to consider the importance of cooperation, as well as competition, as a selective pressure in the origins of social life (Sussman and Chapman 2004). But overwhelmingly they called for the collection of specific kinds of data that would enable the hypothesis to be thoroughly and robustly tested.

Conclusion

This call for particular kinds of data provides the first clue to the reasons behind both the emergence of the infanticide controversy and its intractability. Despite the suggestions by participants on both sides of the controversy that the source of dissension must be sought in factors outside science—in the political applications of biology that have historically marred Western society, in the intellectual ascendancy enjoyed by elite institutions in United States society, in the moral abhorrence of an act as culturally unacceptable as infanticide—other more complex and more subtle influences are also at work here. Paramount among these is the nature of the field environment itself, closely followed by the unique cultural status granted to the nonhuman primates in Western societies. As the following chapters will show, doing science in the field is very different from doing science in the laboratory. It is difficult work, physically and emotionally demanding, requiring the development of a range of techniques, skills, and abilities not normally evoked by the laboratory environment. Where the subject of study is the "natural" behavior of animals as complex, fast-moving, and endangered as the nonhuman primates, deciding what to study and how to investigate it is more difficult still. Each of these factors has a role in creating what was earlier called the fieldworker's regress.

Throughout the controversy, what could constitute naturalistic, representative observation was at the heart of the problem. In relation to the individual field site, questions of who saw what, and how much was seen of a given interaction, recur constantly in the literature, and the ability to surround a particular observation with detailed, personally witnessed context was an essential element in its acceptance as a reliable, authoritative account of animal behavior. The circumstances of the particular field site were also at issue: To what extent were the behaviors seen there typical of a given species, and did this particular place reflect the "natural" circumstances in which that species had evolved? The very processes of industrialization and globalization that had made it possible for primatologists to travel to the faraway field were also influencing the behaviors they had come to study though habitat destruction, tourism, and urbanization. To what extent could any behavior seen in the latter half of the twentieth century be treated as not inflected by human influence? And finally, that the continued existence of many primate species was and is threatened in the wild, in combination with the methodological commitment to the study of the natural behavior of free-living primates, as well as the status accorded to monkeys in Hindu belief made it very difficult for primatologists to contemplate experimental manipulations to test the competing hypotheses: pragmatic, practical, and ethical considerations forbade it.

As part 1 will show, the evolution of the controversy surrounding primate infanticide was closely tied to the historical development of primatology as an institutional and academic specialty. Chapters 1 and 2 will outline the changing approaches to theory and methodology that emerged as field site research expanded in the latter half of the twentieth century and the mutable way theories and conceptions of evolution were themselves apprehended and applied, publicly and privately, over the period in question. Part 2 will deal with the controversy itself. Chapter 3 will describe the initial attempts to study langur behavior and the very different accounts of langur society provided by observers even in the early years of primatological fieldwork. The observations, narratives, conferences, and publications that marked each stage of its development, along with participants' attempts to account for the controversy's existence in the context of the development of primatological field practice, will be analyzed in chapters 4 and 5. Part 3 will turn to the explanations for the controversy's emergence and persistence: chapter 6 will concentrate on the accounts given by those who participated in the controversy, and chapter 7 will consider these accounts in relation to the particular context of the field environment. I will return to the question of the fieldworker's regress in the conclusion.

First, however, I will establish the context for controversy by examining the origins of primatological field studies, the results of the early attempts to account for the existence and structures of primate society, and in particular the studies of langur behavior made before Hrdy's work at Abu. Just as the observations of primate behavior needed to be placed in their historical, social, and ecological context to be accepted as authoritative and systematic accounts of primate life, so the infanticide controversy needs to be considered in the light of the circumstances in which it first emerged—the urgent call for sustained and reliable scientific studies of the behavior of humanity's closest relatives, studies that would provide a sound empirical basis for speculating about human origins.

PART ONE

Fielding the Question

ONE

Primates in the Field: Doing Field Science, 1929–74

Introduction

In the opening decades of the twentieth century, two key figures had turned their attention to what was then known about the bodies, minds, and behavior of the nonhuman primates and the uses this information had been put to. On either side of the Atlantic, Robert Yerkes and Solly Zuckerman[1] were appalled to find that the literature on the lives of wild primates was largely inaccurate, unreliable, and extremely limited and were scathing in their condemnation of their predecessors. For example, in 1929, Robert Yerkes and his wife Ada published their summary of the extant knowledge of the great apes. Even though their review specifically excluded such obvious suspects as "naturalistic or collecting expeditions and . . . transient or long continued but usually somewhat sporadic observations of travellers, hunters and animal trainers and caretakers," they still found that "superstitions, surmises, rumours, accidental and unverifiable observations, inferences and unwarranted conclusions, have been repeated through the centuries" (Yerkes and Yerkes 1929, 582, 2). Solly Zuckerman, writing in 1932, complained that "information about mammalian social behaviour has continued to accumulate mainly in the form of travellers' tales—tales which seldom rest upon the accurate personal observation of their narrators; anecdotes in which factual and interpretive elements are inextricably mingled" (Zuckerman 1981/1932, 11). Although both books argued that studying the primates, particularly their social behavior, would present a wonderful opportunity for developing "experimental sociology and important new departures in social psychology" (Yerkes and Yerkes 1929, 255), they complained that the "web of romance" (Zuckerman 1981/1932, 2) in which the subject was ensnared meant such an endeavor was impossible at that time.

Knowledge about the nonhuman primates, especially the apes, was important, they argued, because these animals could act as substitutes for the bodies and minds of human beings. This had worked well in the laboratory, where apes and monkeys had been used as research subjects in anatomical, physiological, and psychological studies,[2] but when it came to studying their social or group behavior, it appeared that speculation was the order of the day.[3] In an early example of the kind of extrascientific uses the primatological literature had been and would be put to, Yerkes and Zuckerman suggested that writers found themselves in the happy position of being able to pick and choose amid accounts of nonhuman primate society.[4] The literature available was so incomplete, so partial, and so contradictory that it could be taken to support almost any position one might choose to adopt. But how could more certain knowledge about the nonhuman primates be achieved? For Yerkes in particular, natural social behavior could not be understood in the artificial conditions of the laboratory: only in the field could one study it in its original context, the conditions under which these behaviors had probably evolved and that might also have applied to the evolution of human behavior.[5] But how should one go about studying behavior in the field?

This chapter will explore the origins of the field research tradition in behavioral primatology, concentrating on the methodological issues of central concern in the debates running through the discipline during the first decade and a half of modern primatological field research.[6] Essentially, researchers were trying to establish how one might go about the systematic, scientific observation of natural behavior under field conditions. This raised a number of interrelated issues—issues that often could not be resolved but had to be constantly managed at the levels of both the individual and the discipline as a whole. Since researchers wanted to study animals living under natural conditions, they had to travel to the field. Given this, at least two further sets of questions were immediately posed. How could one do science in an environment characterized by conditions so very different from those found in the laboratory, the iconic scientific space of the twentieth century? And what was meant by this innocuous term "natural conditions"? But these were not the only problems nascent primatologists faced: for example, people working on primate behavior came from many nations and from diverse backgrounds, thus ensuring that the difficulty of translating between languages and disciplines also affected these debates.[7] As this chapter will show, however, it was in the attempts to answer these questions that the roots of the infanticide controversy grew, as researchers developed pragmatic compromises and partial solutions to intractable problems such as the definition of nature, the adjudication of humans' place in the natural world, and the need to standardize the (almost) infinite variety of places and social structures where primates could be found. At the same time as they were observing the behavior of free-living primates, researchers were being forced to make decisions on *how* such observations should be accomplished so they might be transmitted to interested communities as accurate reflections of some aspect of the natural world. It was under these conditions that the "fieldworker's regress" was to flourish.

Prewar and Postwar Primates

Robert Yerkes provided the impetus for the first attempts at naturalistic, systematic study of the great apes to establish factual knowledge that could "imply more than hunter, collector, or wandering field naturalist can supply" (587).[8] In pursuit of this project, he sent three students to observe the behavior of chimpanzees (Henry Nissen), gorillas (Harold Bingham), and howler monkeys (Clarence Ray Carpenter). Of these three attempts to study primates in the wild, Carpenter's was undeniably the most successful, and in 1934 he published his account of the social life of the howler monkeys of Barro Colorado Island in the Panama Canal. Based on eight months of observation during 1931–33, Carpenter regaled his audience with details about not just how and where the animals moved and what they ate, but the way their groups were integrated, social relations within and between groups, how group behavior was coordinated, the relationship between the sexes, and the ways both sexes responded to infants and juveniles.[9] In his foreword to this monograph Yerkes concluded that this not only was the "first reasonably reliable working analysis of the constitutions of social groups in the infra-human primates" (Yerkes 1964, 4)[10] but also provided a standard against which successors could measure themselves, most especially Carpenter's analysis of the practical problems of observing primates in the field and the methodological solutions he suggested.[11]

However, this auspicious beginning did not mark the onset of the systematic studies that Yerkes had so eagerly anticipated. Although laboratory research on the anatomy and psychology of the primates continued, as did the attempts to theorize the nature and origin of primate sociality,[12] the field-based reports published in the 1940s and 1950s tended, with a few exceptions, to be either based on isolated observations or undertaken as an adjunct to other research interests. So, for example, in 1940, Beatty reported seeing a chimpanzee cracking nuts in Liberia (Beatty 1951); in 1953 Nolte made a brief examination of the behavior of Indian macaques (Nolte 1955); the East African Virus Research Institute had made fairly detailed descriptions of the behavior of free-living African redtail monkeys (Haddow 1952) as a complement to its laboratory work; and March made unsuccessful attempts to watch gorillas in Nigeria during 1955 and 1956 (March 1957). More systematic studies of the howlers of Barro Colorado had been conducted by Collias and Southwick in 1951 (Collias and Southwick 1952) and by Altmann in 1955 (Altmann 1959). Bolwig had watched baboons in South Africa in 1955 (Bolwig 1959b), and the prosimians of Madagascar had been studied in 1954 and 1956 by Jean-Jacques Petter, among others (Petter 1965). And finally, although largely unknown to Western researchers until the late 1950s, Japanese researchers had been studying the behavior of Japan's indigenous macaques since 1948 (Frisch 1959). This lack of prolonged, naturalistic studies of nonhuman primates, such as those provided by Carpenter, was deeply regretted by anthropologists and anatomists like Ernest Hooton and Adolph Schultz. During an interdisciplinary symposium on primate studies held in the mid-1950s, Hooton made his views plain: "I view with utmost dismay the lack of sustained interest in such studies, the present reluctance of institutions to promote them and the difficulty of financing them" (Hooton 1954, 186). Having reviewed the accomplishments of primatology in the first half of the 1950s, Schultz regretfully concluded that in fieldwork, "little progress has been made in this direction since the exemplary and well-known contributions by Carpenter" (Schultz 1955, 55). But this situation was soon to change.

Less than a decade after Schultz complained of the moribund state of behavioral primate research, Irven DeVore and Richard Lee (1963, 67) were able to publish an (in)complete list of completed, planned, and ongoing field research on primates that ran to nearly fifty entries.[13] The modern period of primate field research in Europe and North America really began in 1958–63. For apes, this was marked by the establishment of the Gorilla Research Unit in Uganda, supported by the University of the Witwatersrand (Dart 1960, 1961; Emlen 1960; Tobias 1961; Schaller 1965a, 1965b); the three separate studies of chimpanzees that began in the early sixties (Goodall 1962; Reynolds 1963; Kortlandt 1962); and Schaller's brief observations of orangutans on Sarawak in 1960–61 (1961). Hall's baboon studies began in South Africa in 1958 (1960), the work of Washburn and DeVore on the baboons of Kenya and Rhodesia began in 1959 (1961), and Kummer and Kurt published the results of their work in Ethiopia in 1963. Surveys of rhesus macaques in India by Southwick, Beg, and Siddiqi were conducted in 1959 and 1960 (1961a, 1961b), and Simonds made behavioral observations in 1961 and 1962 (1965). Jay had watched Indian langurs from 1958 to 1959 (1962), and the Japan India Joint Project in Primates Investigation had done so from 1961 to 1963 (Sugiyama 1964). In addition, the Japanese Monkey Centre, itself established in 1956, had sent out teams in 1958 to survey Africa and Southeast Asia for sites where wild primates could be watched, to complement their work on captive and provisioned Japanese macaques (Frisch 1959). Concentrated overwhelmingly on apes and large-bodied, mostly terrestrial monkeys in Africa and Asia, these projects represented the first wave of primatological field site studies, studies that were to grow exponentially over the next four decades as primatology became institutionalized as an independent discipline.

The emergence of societies, symposia, and dedicated journals marked the onset of this process of academic institutionalizing in the early 1960s, with behavioral studies of free-living primates taking center stage in the development of this new discipline. For example, in 1962 alone four meetings were held in Europe and the United States. Papers dealing with the behavior of both wild and captive primates dominated the British and American meetings (Schultz 1964), although the continuing importance of the use of primates in biomedical research was illustrated in the fourth conference, held at Beaverton, Oregon, to mark the opening of a new primate research center there.[14] In 1963 the International Primatological Society was founded, and its first congress was held in Frankfurt in 1966 (Carpenter 1969). Also in 1963, the first Western journal dedicated to the study of the primates appeared (the Japanese journal *Primates* had been founded in 1959, although it did not attract much attention in the West until some time after it began to publish in English).[15] *Folia Primatologica,* on the other hand, quickly began to publish the results of Westerners' field studies of free-living primates. In fact, the very first article in this new international journal was Kummer and Kurt's account of their year in the field studying the baboons of Ethiopia. Seven others were to appear over the next four years,[16] and field studies continued to be reported in other journals such as *Behaviour, American Journal of Physical Anthropology, Science, Man,* and *Journal of the Zoological Society of London.* But the appearance of *Folia Primatologica* is interesting not just because of its significance as the first Western journal dedicated to this emerging discipline, but also because of the explanation given for its origins. The editors explicitly located its genesis in Huxley's admonition to "look at man's place in nature": on this reading, the natural origins of humanity were to be the "foundation of modern primatology" (Anon. 1963, 1). This both continued and endorsed a tradition that had a long history: as Yerkes and Zuckerman had pointed out, the nonhuman primates were of interest because of their similarities to human beings and their capacity to substitute for humans. Like primates in the laboratory, free-living primates were ultimately to be used as a means of studying some aspect of human life inaccessible in modern human bodies and minds, for either practical or ethical reasons.

Establishing Field Traditions

Even as this new disciplinary specialty established its institutions, so the range of primate species studied in a number of locations grew from the late 1950s into the mid-1970s.[17] Researchers, however, continued to focus on the problems and questions Carpenter and Zuckerman had raised in the previous decades. These issues not only related to the growing body of knowledge about the lives and behavior of nonhuman primates, which I will discuss in detail in chapter 2, but also persistently interrogated the contexts within which this information about the lives of the primates is obtained, and the uses it is, or can be, put to. It is evident from the concerns researchers raised in this period that Yerkes's vision of the establishment of sites where "work can continue uninterruptedly for years and where observations on mode of life and environmental relations may be checked, verified and supplemented as desired" (Yerkes and Yerkes 1929, 587) represented an ideal that proved extremely hard to realize.[18] Researchers continued to emphasize the unsatisfactory nature of their knowledge of primate behavior: in particular the lack of systematic, reliable studies of social behavior in the wild as opposed to the analysis of dyadic interactions in the laboratory.

This dissatisfaction can be traced back to the challenge Carpenter laid down in 1942, when he called for the observation of

> whole animals in natural, organised, undisturbed groups living in that environment which operated selectively on the species and to which the species is fittingly adapted . . . [and the application of] the scientific method in field studies of non-human primates. . . . Absolute objectivity, accuracy of recording and report, and adequate samplings of observations can be made to characterise alike field investigations and those of the laboratory. (Carpenter 1964, 342–43)

Two themes that were to prove crucial to the development of primatological field practice are being drawn on here. The first is the assertion that that the proper object of the primatological gaze is not just the behavior of the individual animal but its behavior in the context of the whole group, which in turn must be observed in the environment in which the species evolved. Only in this way can the adaptive origins of behavior be understood. This approach was to become characteristic of many examples of field science, emphasizing the holistic nature of the topic of study: animals' actions cannot be understood in isolation from their social and environmental context. Systems must be studied in interaction, not in isolation, as in the laboratory. Carpenter's second assertion, however, explicitly rejected the notion that the study of behavior in the field need be any less "scientific" than its study in the lab: objectivity, accuracy, and the description of representative behavior are to be sought and achieved in both locations.

His need to affirm the potential equivalence of these two spaces is revealing: as he admitted in the same article, "The fashions of science since about 1900 have so strongly favoured ever increasing laboratory control that there has been serious neglect in applying the scientific method to field studies of animals" (Carpenter 1964, 342). The historic focus on the laboratory as the place where the natural world can be controlled, manipulated, and tested and from which standardized, generalized systematic knowledge can emerge, meant that a sharp contrast could be drawn between field-based and lab-based research with respect to the authoritative status that interested parties granted to the knowledge emerging from these distinctive places. Field research, in this sense, was carried out in an open, uncontrolled environment, replete with variability and detailed particularity, and apparently antithetical to the ostensibly pristine laboratory sites characteristic of modern science. As this chapter—indeed, this book—will demonstrate, working within the field environment demands far more interpretive flexibility from the researcher than is called for in the laboratory. Carpenter's initial assertion that it was possible to create authoritative field-based knowledge of primate behavior that could stand comparison with that produced within the laboratory was made at a point where systematic studies of primates in the wild had not quite begun. The history of primate field science in the latter half of the twentieth century illustrates the problems his successors had in putting this into practice. The progress of the controversy surrounding primate infanticide in particular was to be materially affected by the two themes Carpenter outlined. First, participants in the controversy had to decide whether the animals that made up their research populations were in fact to be found in "natural, organised, undisturbed groups living in that environment which operated selectively on the species and to which the species is fittingly adapted," and second, they had to establish how to achieve accurate reports, adequate sampling, and objectivity in the field context. In particular, what was to be considered a "natural" primate group would prove critical. Interpretive reflexivity was characteristic of primate field site research from the outset.

Observing in the Field

Given the influence Clarence Ray Carpenter had over the development of primatological field research in the United States,[19] it is useful to consider his definition of an authoritative, systematic field report. He emphasized that, to achieve reliable results that could be drawn on to produce what might become general laws of primate behavior, field research had to fulfill a number of conditions. Although field-study methods would inevitably depend on the species being observed and the type of environment it lived in, any study would have to ensure that the *effect of the observer's presence* on the animals' behavior was minimized (Carpenter 1964, 94).[20] As many animals as possible should be watched for the longest period achievable (18), consisting of a *representative* sample of the population found in an *undisturbed* habitat (162, 366). Behavior should be recorded as it occurred and should be written up and classified each night (18, 164). The observations themselves could be accepted only "after confirmation and re-checking and if all details necessary for a reasonable interpretation were known" (18), since field observations "primarily involve description and as such are rather *subjective*" (161; italics added). Clearly, putting these conditions into practice immediately raised further questions. How minimal must the effect of the observer be, and can this effect be measured? What counts as a representative population or an undisturbed habitat? What details are necessary for a reasonable interpretation, and how can the problem of subjectivity in the field be overcome? Not surprisingly, researchers' reports during the 1950s and 1960s demonstrated the range of interpretations that might be made of Carpenter's methodological specifications.

In a later paper[21] Carpenter went further, outlining not just the basic conditions needed for conducting fieldwork but the kind of information that should be included in a field report. Specifically, it must cover topics such as "geographic distribution, ecology, taxonomy and characteristics, population structure and dynamics, behaviour, complex group characteristics and general deductions consisting of abstracted concepts, theories, generalisations and inferences" (1965, 250–51). Evidence must be collected on habitual, not just unusual, events. The duration and amount of observation had to be sufficient to answer the questions asked; observations should be repeated and confirmed, preferably by different observers and with the use of recording equipment, to ensure reliability. Although disturbance was to be minimized, it might sometimes be necessary for the observer to achieve a "skilful" entrance into the group, always bearing in mind the need to adapt methodologies when dealing with different species (1965, 255–56). Ultimately, however, careful attention had to be paid to the observer's personality, since differences in individual researchers' mental and physical capacities have far greater consequences for field research than for laboratory work. As Carpenter argued,

> The collection of evidence in the field depends on the macroscopic observation capacities of the observer. . . . The characteristics of the observer influence results more in naturalistic than in laboratory studies, where observational demands are limited and instruments can be used to greater advantage. . . . The question must be asked: Who made the observations, reports and inferences, and what were his qualifications for making them? (1965, 255–57)[22]

Here Carpenter has created a situation where the authoritative status of the primatological field report during this early period is closely tied to its context of production. The confidence one could place in such an account was related to how detailed a description it provided of the place where the research was conducted, the animals under observation, and the person doing the observing. Unlike reports of research conducted in the laboratory, where including such material would be grounds for questioning the report's reliability, the contexts of personality and place were essential elements in evaluating and interpreting the trustworthiness of these early accounts.[23] Primatologists went to the field to investigate areas of behavior impossible to access in the laboratory, and these field sites were useful precisely because of the characteristics that separated them from laboratory space: they were variable, unpredictable, and complex, but they contained resources that could, with the proper management, provide the basis for formulating general laws out of particular places.

The Laboratory versus the Field?

The relation between the laboratory and the field was a key theme in methodological discussions throughout this period, as researchers repeatedly questioned both the relative status of these two research contexts and how far laboratory techniques might be successfully exported to the field. The case for mutual assistance, and the list of problems that would have to be overcome to achieve successful collaboration were both eloquently described. Carpenter had argued, along with other researchers, that in an ideal world, "field observations would suggest problems for laboratory investigations and would assist students of behaviour to interpret and, in a sense, to validate the results of laboratory research" (Carpenter 1965, 255).[24] In other words, the two sites of knowledge production could and should work in tandem. Laboratory researchers, for example, could experimentally investigate the physiological causes and consequences of a given behavior in a particular animal or group of animals, happy in the knowledge that there existed an unmolested "control" group of animals living under wholly natural conditions, to which their results might be fruitfully compared.

However, though some researchers did cross from laboratory to field and back again, the chasm between the two yawned. The most important reason for this gulf was the imbalance in the authoritative status granted to the knowledge thought to emerge from these different locations. In a sense this was inevitable, given the tensions between the definitions of experimental and observational purity or authority used in laboratory work and in fieldwork. Where one site stressed its ability to control the actions of different variables and the other its reluctance to intervene in the normal progress of events, questions could arise as to who was obtaining the more accurate, or possibly the more relevant, results. So, for example, William Mason pointed out that "the experimentalist . . . is inclined to doubt that the fieldworker, unaided by the instruments and controls of the laboratory, can really achieve an objective and reliable description of behavior," while the fieldworker, "knowing something of the artificiality of the laboratory and the constraints it imposes on the animal . . . is likely to question the value of experimental findings" (1968, 401). The danger was that the animals were studied under such radically different conditions that researchers in either environment doubted the reliability of data obtained in the other. And yet it always seemed that fieldwork was particularly vulnerable to criticism.[25]

Ironically, as Plutchik (1964) pointed out, many of the problems Carpenter identified as especially significant for the categorizing and sampling of social behavior in the field also applied to the laboratory but needed to be solved in rather different ways. For example, as early as 1950, Theodore Schneirla, an animal psychologist,[26] had echoed Carpenter's 1942 criticism of the "mistaken impression that control may be obtained only under laboratory conditions" and had pointed out that "observation and experiment are as closely related in field study as they are in other areas of scientific investigation" (1950, 1022). Both places "include perceptual observation in some form as an essential procedure, and the fundamental criteria of reliability and validity of evidence are similarly involved, as are techniques of control" (1024). Crucially, however, the difficulty of eradicating this "mistaken impression" is demonstrated in that eighteen years later Mason had to emphasize that there is still "no evidence for a fundamental difference between experimental and field research. Whether the investigator works in the laboratory or in the field, his basic aim is to achieve a reliable, objective, description of behaviour. In either setting, observer bias and the effects of the observer's presence on the objects of study must be evaluated" (1968, 400).

Part of the problem for Mason, among others, seems to have been the different personalities that were considered characteristic of fieldworkers and laboratory workers. As Carpenter had suggested, it was felt that the personality of the fieldworker could potentially have a profound impact on the development of a field study. Hall had pointed out in 1965 that fieldwork requires "an unusual combination of intellectual and temperamental characteristics, which will enable him or her to persevere, often in harsh climactic conditions day after day, often for ten or twelve hours a day, with few, if any social or scientific amenities" (1968a, 10). This position was echoed by Mason, who argued that the fieldworker "must be able to accept solitude, physical discomfort, and frustration as normal parts of his daily routine" (1968, 401), and by Carpenter (1965, 256–57). Exceptional talents are required, and Carpenter actually suggested that fieldworkers themselves "deserve to be studied to determine their unique characteristics [so that] both our normal and deviate personality characteristics, in some manner, [can] be screened and prevented from distorting the objectivity and emphasis of observations and their recording and reporting" (1962, 492). What seemed certain was that the personality traits that would lead to success in the field were very different from those required in the laboratory, and as Schneirla pointed out, it was worrying that "different observers often obtain very different results from what is objectively the same phenomenon . . . in studies of social behaviour, our attitudes toward the nature, origin and relationships of competition, cooperation and natural selection processes exert subtle influences not readily controlled in the planning and prosecution of investigations" (1950, 1028).

The question, in these accounts, then became how to exert control in the field environment. Fundamentally, as the next section will show, doing science in the field became a matter of managing not just space, but also self. Not only must one ensure that the population was not disturbed unnecessarily, but one must constantly monitor both one's perceptions and the use of equipment to extend the range and depth of sense perceptions (binoculars, cameras, tape recorders). The procedures of the laboratory continued to function as a *conceptual* resource—so that, for example, just as experiments can potentially be replicated, so can observations. The single observation represented only the starting point in the analysis of behavior, which had to be repeated in various social situations and environments before it could be accepted and used in formulating predictive hypotheses. However, in developing field-based methodologies, fieldworkers did not just draw on the experiences of the laboratory; they also adopted the social practices of other inhabitants of the field site, both human and nonhuman, to find ways of studying and accounting for the behavior of the nonhuman primates they watched.

Developing Field Methodologies

The fact that field sites, unlike labs, do not operate as physically delimited spaces meant researchers found that their potential research sites frequently were inhabited by miscellaneous groups of people, including but not limited to farmers, poachers, tourists, hunters, loggers, officials, and servants. The presence at "their" sites of such a variety of social groups had consequences for the reception given to their reports and the strong perception that the field lacked the laboratory's authoritative status. For example, regarding field biology in the early part of the twentieth century, Kohler points out that the "social diversity [of the field] also compromised field biologists' credibility and social standing" (2002a), and Kuklick and Kohler have noted that "the margins of settled Western society have been places of refuge for deviants. . . . Naturalists have been mistaken for bandits or ne'er-do-wells because people of uncertain standing have found the margins to be congenial places" (1996, 11). But as they go on to point out, "field scientists are more likely than laboratory scientists to suffer (or *benefit from*) uncertain identity" (italics added). While the gentlemen of the seventeenth century drew on their own culture to form the basis for the customs and practices of the laboratory,[27] the presence in the field of so many disparate groups has provided the field scientist with a number of models, such as administrators, sportsmen, farmers, missionaries, mystics, and travel writers.[28] Notably, primatologists have adopted some of the social practices used by these other groups as part of forging a field identity for themselves.[29] As I will show, the basic research method for behavioral primatology became the all-day hike, following the selected group or individual; researchers camped in forests and on mountains and open plains, and they trailed their animals using techniques originally developed to hunt them. Like tourists, initially they followed and observed their animals from Land Rovers, and in some cases, to habituate the animals, they borrowed social techniques from the animals themselves, grunting like an incoming conspecific[30] to avoid startling their study group, eating the foods the primates favored, and imitating their strategies for appeasing an irate dominant. These social and practical skills all played an essential part in allowing researchers to solve the problems of working with wild-living animals (Rees 2006a).

The fundamental difficulty was to create a position from which the animals could be watched regularly and reliably. The principles of observation Carpenter had outlined were clear: as many animals as possible were to be observed for as long as possible, with observations being repeated and confirmed, preferably by different researchers, so as to reduce the subjective element in observational interpretation. Obviously, putting these prescriptions into practice depended on how visible the animals were in the first place. This was why most of the early studies in modern primatological field research focused on relatively large animals that spent most of their time on the ground: it was far harder to see animals in the trees.[31] But no matter their preferred substrate, the initial response of most primates to the novel presence of human beings was fight or flight. So, for example, Carpenter reported that howler males would roar at him (1964, 20), an observation Donisthorpe echoed several decades later for gorillas (1958, 214), and Schaller described the way orangutans annoyed by his presence would drop branches, causing him "to jump nimbly at times and [keeping him] effectively away" (1961, 82). Booth, who took a tamed infant monkey with her while observing wild troops, found that the animals became extremely agitated by the infant; one male approached so closely that "it was not only possible but expedient to discourage him by hitting him with my hand" (1962, 484). Hall's study of several baboon troops at the Cape of Good Hope Nature Reserve found that they varied widely in their responses to human presence: some troops were tolerant, others would run away. DeVore and Washburn's study of baboons in Kenya had to be conducted from a Land Rover because the baboons, although used to such tourist vehicles, became nervous when confronted with humans on foot. The aggression, or fear, that the animals showed to a strange observer did not just make it difficult to watch them; it reflected active disruption: behavior provoked by the presence of a stranger was clearly not "normal." The first goal of a successful field study was thus to somehow accustom the animals to an observing human.

In other words, the animals had to be *habituated* to the observer's presence, and researchers used two main methods to achieve a social position within or near the group from which they could watch without apparent disturbance. They could approach the animals slowly, stopping when they displayed agitation and hoping that this "flight distance" would steadily diminish until they were close enough to see what was going on within the group. Or they could tempt the animals with food, a technique first used by researchers from the Japanese Monkey Centre, or JMC (Imanishi 1960, 393; Miyadi 1964).[32] With this method the study of social behavior could proceed much more quickly, but as Frisch warned, when abundant food is presented to a group, the danger is that "some items of social behavior [are] modified or suppressed, or perhaps intensified by the new situation" (1959, 594).[33] In most cases the technique of slow approach was used, supplemented initially by observations made from hiding (Altmann 1959; Emlen 1960; Hall 1960; Kortlandt 1962), but always with the awareness that the observer's presence risked changing the animals' behavior.[34] Emlen and Schaller, in their observations of gorillas in Uganda, found that when the animals became aware of them, they would come together to "watch the observer from an open situation at favourable range," a response that frustrated their "ultimate objective of studying undisturbed animals" (1960, 89). But as the animals became used to observers, it appeared that their behavior returned to "normal." For example, DeVore found that when their observations of the baboons around their car were compared with "observations made at a distance (through binoculars) both at Amboseli and in Nairobi Park, it was clear that the presence of human observers in an automobile made very little difference in the social behavior of these semi-tame troops" (1963, 307). Hall perhaps best sums up this methodological compromise when he concludes that "the mere continued presence of a "neutral" observer in a setting where the animals are not otherwise usually frightened by human beings, is sufficient to ensure the conditions necessary for close-range field study" (1963, 22).[35]

Habituation was accepted as a means of familiarizing the two sets of participants in the field study with each other without undue disturbance, but this technique had profound implications for what was to count as a representative population of primates. Not only was it difficult and initially unrewarding work—as Carpenter pointed out ruefully, the field-worker needed "the endurance and patience of a pack mule" (1965, 257) to apply it—but it seemed simply impossible to use in some situations. Animals that were arboreal, or lived in the forest edges where visibility was extremely poor, or had been hunted and were hard to habituate. It was a method that applied best to the large-bodied terrestrial primates such as baboons or the semiterrestrial langurs, and to the social, forest-living great apes. Thus, as participants realized, it produced a bias in the early descriptions of nonhuman primate social life. Additionally, it was a method that worked most swiftly and successfully in a situation where animals already had a relationship with human beings. The question, however, was whether previous exposure to humanity meant the primate population in question had been "disturbed." For two reasons then, using habituation as a methodological technique had profound consequences for the early years of primate field studies: it meant that observations were concentrated on those species that were easiest to see, and to be seen by, and on those groups that already had a history of contact with humanity.[36]

The Importance of Place

Carpenter's emphasis on watching undisturbed and representative populations therefore needs to be examined rather more closely in relation to the selection of sites to work at and animals to watch. The notion of watching an undisturbed population can be taken in two distinct ways. First, it can mean an untouched population, in the sense that the animals and their environment have both remained relatively unaffected by human environmental alterations (farming, logging, mining, and so on). Second, as the previous section showed, it has implications for the observers' behavior: most centrally, that they will not intervene in the animals' activities. In either case, the fundamental point is that the animals' behavior should not have been contaminated by human influence: the best kind of study is carried out in a place where the nonhuman primates are thought to be relatively unaffected by human activities.[37] On the first point, in practice most fieldworkers, even in this earlier period of research, found themselves working at sites that fell somewhere along a continuum of interference, depending on the species of the primates, their habitat, and the attitude of the local human population. Populations that were undisturbed would by definition also be very hard to locate and get to, and most of the early studies were conducted on primates that were easy to locate.[38]

The continuing importance of this issue is apparent in that, while the contexts of the early studies differed, research reports consistently defended the sites' status as undisturbed environments. In particular, note that the semisacred cultural status of nonhuman primates in India specifically and Asia more widely meant that defining their "natural" habitat as excluding human presence was very hard to do (Carpenter 1964; Southwick, Beg, and

Siddiqi 1961a, 1961b), and in fact, the preface to DeVore's edited exemplar of primate studies opens with the assertion that "man's acquaintance with the monkeys and apes is as ancient as man himself" (DeVore 1965, vii). So, for example, Southwick and his colleagues make the point in relation to their studies of the rhesus macaque in India that it is hard to determine "what the natural habitat of the rhesus actually is. It has lived in close ecological contact with man for centuries, and in frequency and persistence, this commensal relationship in villages, towns, temples and roadsides represents a natural relationship" (Southwick, Beg, and Siddiqi 1965, 158). The impact and intensity of this "natural relationship" could be seen in the sharp decline of rhesus numbers where cultural attitudes toward monkeys had changed and where environmental conditions had intensified food competition between the human and nonhuman primate populations.

Close ecological, if not social, relationships between human and nonhuman primates were characteristic of, but not restricted to, the Indian subcontinent. For example, take some of the field studies carried out in Africa during this early period, specifically the baboon populations studied by Hall, Washburn, and DeVore, which were frequently in national parks supported by tourist revenues. While watching patas monkeys, Hall was forced to change his study location, partly because "the patas, frequently raiding crops, were periodically harried and hunted by the human inhabitants" of the area; for him, the greatest advantage of the new study site was that "there would be no interference from human[s]" (1968c, 44). Naturally this raised the question of the conservation of endangered primate species very early in primatology's history: for example, Dart, Emlen, and Schaller question the future of the gorilla, given human pressure in the Ugandan/Rwandan area and the growth of nationalist movements, and Schaller (1961), Thorington and Groves (1970), and Southwick, Siddiqi, and Siddiqi (1970) cite hunting, habitat destruction, and the trapping of animals for biomedical research as the key features affecting their natural behavior at present and threatening their future. However, the relationship between human and nonhuman primates at the places where foreigners chose to work was not inevitably damaging. As Southwick found, working with Indian colleagues, in the right circumstances local people could be supportive of the primate populations, if not necessarily of the research in progress.[39] In fact they suggest that in some senses macaques could be said to "exist as 'public property' within the public domain" (Southwick, Beg, and Siddiqi 1965, 132). Again, particularly in India, local people doubted the motives of foreign researchers and suspected they had designs on "their" primates. For example, Sarah Hrdy described being attacked by an elderly man who had seen her squirt hair dye on some langurs (for identification) and assumed she was marking them for slaughter (Hrdy 1977a).[40] The perceptions and opinions of local civilians and authorities had to considered before research could be undertaken at a particular place (Rees 2006a).

Let me raise a final point about this notion of the natural, undisturbed, representative field site. Clearly, while primatologists acknowledged this as the ideal, in practice most worked in places that could not measure up to this standard. In addition, however, this concept of the ideal site was profoundly static. In reality, primatologists increasingly found that conditions at the sites where they worked were in flux, sometimes because of human influence, sometimes not. In fact, Rowell had pointed out in 1967 that in practice fieldworkers knew little about the history of the animals at their research sites and tended to assume that the site was "stable in its present state" (Rowell 1967, 223). This was not a safe supposition during a period of intensive rural and urban development, closely linked with rapid environmental change. Jay acknowledged that whereas in the past "emphasis almost had to be on locating study areas where animals occupied an environment as similar as possible to the one in which they lived before man altered much of the land's surface" (1968, 174), this situation had had to change. While researchers still wanted "to observe in as unaltered a situation as possible [they now realized] that some species are able to thrive in many different habitats, some of them much modified by human activity" (Dolhinow 1972, 19–20). Studies at such sites were valuable precisely because they demonstrated primates' behavioral plasticity—they were capable of adapting to different environmental conditions.

Additionally, it was becoming progressively more evident that long-term research at particular field sites would be vital to understanding primate social behavior. It became clear, not least from the observations of the Japanese Monkey Centre researchers, that as the time spent studying a group of primates increased, the quality of the information obtained changed. Details became available on the animals' genealogies and on their individual life histories, enabling a contextualizing of their behavior that had a major impact on the interpretations of the observations (Frisch 1959; Imanishi 1960; Simonds 1962), and researchers like Dart (1961), Hall (1963), Mason (1968), and Rowell (1966) all emphasized the role that long-term observations would play in understanding the way behavior, even in the apparently well-known terrestrial species, might be found to vary with environment. For these reasons it was even more important for researchers to include a careful description and discussion of the particular environmental, ecological, and social conditions at their sites and their advantages and disadvantages for the purposes of the investigation, as an essential aspect of their reports on the behaviors observed there. As Carpenter had suggested, the context of the field report was central to evaluating its content.

Experimenting with the Field

Historical change afforded the field primatologist another set of opportunities: taking advantage of "natural" experiments. These could occur in a number of forms but usually meant either unexpected variations in the ecology of a known troop or the discovery of animals living in unexpected environmental or social conditions. So, for Collias and Southwick, the sharp population decline on Barro Colorado Island[41] represented a "natural experiment in which we were able to inquire to the effects on social organisation in a natural population of halving the density of that population" (1952, 145–46), while Hall was able to observe the effect that drastic food shortages had had on the baboon population at Kariba, Rhodesia (1963). He was also able to compare the organization of the baboon troops at this site with those at three other sites in southern Africa in order to demonstrate the interplay between diet and social complexity and, more generally, to show the significance that more controlled variations in captive groups might have for understanding behavior. More broadly, the island of Madagascar itself represented an "unique natural experiment" for Bourlière (1968, 648), because it offered an opportunity to see how prosimians had evolved in the absence of monkeys. Similar strategies were to be used in relation to the almost equally unfamiliar New World primates. But *deliberate* experimental manipulation in this period is notable for being discussed with enthusiasm but practiced only hesitantly.

Researchers hoped that experiments could "be designed to create precisely those situations needed to settle uncertainties as to which elements are important in an interaction or situation" (Jay 1968, 377), and some attempts were made to manipulate certain areas of the lives of the animals under observation. For example, despite the concerns about the impact that provisioning might have on the behavior of free-living primates, many researchers used food tests to determine the dominance relations within a group (Hall and DeVore 1965).[42] Hall also showed animals models of dangerous creatures such as snakes or cats to monitor their reactions (Hall 1960, 1962) and, in a similar vein, exposed a group of patas monkeys to a captive juvenile monkey on a leash (1968c, 104). Other attempts at manipulating social structure had been made by Carpenter and also by Kummer, who used translocation experiments[43] to study the basis for the harem groups seen in hamadryas baboons (1968). The importance of developing techniques for field experiments was constantly emphasized, but researchers were clearly wary that, as Hall put it, "there is far too little control of the relevant factors" (1968d, 391).[44]

Instead, researchers turned to quantifying data and devising sampling methods as a means of developing archives that would not just be a representative account of behavior observed at a particular site but could be used in cross-site comparisons—a technique that was to become fundamental to the progress of the infanticide controversy. An early example was Altmann's attempt to define, quantify, and statistically interpret the behavior of the Cayo Santiago macaques (Altmann 1962a). Another was Hall's call for quantifying observation as the essential first step in developing reliable, objective knowledge about primate behavior in the field. Verbal, as opposed to numerical, statements were "quite inadequate scientifically" (Hall 1963, 1968b, 122). Thus it was imperative that "first and foremost, field observers should be trained to think quantitatively, for this will affect their whole approach to the task" (1968b, 123). Such data, presented numerically, would enable cross-site comparisons—a potential substitute for the replication essential to the accounts of laboratory research—and might suggest avenues for developing more sophisticated interventionist, or experimental, methods for the field.[45] By the late 1960s, Stuart Altmann was able to assert that for some primate species at least, field research had moved away from simple description and toward the more exact analysis of behavior based on the use of quantitative methods (1967, xi–xii).

But a further step was necessary: it was not enough to count what was happening. One had to decide what one was counting and why. In 1974 Jeanne Altmann published a paper in *Behaviour* that became one of the most influential methodological surveys not just in primatology, but in animal behavior more generally (Haraway 1990). She pointed out that the observer "often chooses a sampling procedure without being aware that he is making a choice" (Altmann 1974, 229) and that the ad libitum data emerging from sites, presented as what Altmann calls "typical field notes" (235), were scientifically inadequate, even when offered in quantitative form. She pointed out that "without some form of systematic sampling procedure, there appears to be no way to avoid the bias that results when the observer's attention is attracted by certain types of behaviour or certain classes of individuals" (237).[46] Her paper analyzed a number of sampling methods, providing a key route through which primatology was able to progress from descriptive natural history to scientific analysis, *without* necessarily resorting to experimental manipulation. Her elaboration and analysis of sampling methods allows for the production of data "through means which are non-manipulative and are therefore less likely to alter or destroy the social system that is being studied" (1974, 231) but that could nonetheless be treated as reliable and valid representations of the natural world.[47] By the early 1970s it appeared that primatologists had reached a tentative consensus on the appropriate methods for studying primate behavior in the wild.

Conclusion

Over this period research in behavioral primatology aimed to replace the anecdotal, anthropomorphic, and sporadic accounts that had dominated the literature at the beginning of the century with systematic, representative studies of the undisturbed behavior of animals living in natural groups within a complex ecological setting. The difficulty, as this chapter has shown, was to decide how such studies might be produced and how to relate them to laboratory studies. Data produced at a field site seemed to have lower status than data produced in the laboratory and to be considered less reliable: unlike the lab, at the field site events could not be controlled without damaging or destroying the behavior observers had come to watch. In addition, how far results emanating from one particular site could be taken as representative of the behavior of that species in general was still being explored. In summary, primatologists found they had to learn to manage a number of factors concerning field site research, not least their own relation to the other primates, both human and nonhuman, who inhabited that space, as well as the way they presented their site to the outside world. The idealized version of a field site where natural behavior could be observed systematically had to be mediated through the practical problems of doing work in the field and in a particular place.

Wary of the costs as well as the benefits of their multidisciplinary origins, primatologists attempted to acknowledge methodological questions and tensions and discuss them in detail. In some ways one could see this attempt as a strategy for avoiding just such a controversy as grew up around the topic of infanticide—an implicit embrace of the fieldworker's regress. The nascent methodological consensus that had emerged from the first part of the second wave of primate field studies in the twentieth century revolved around techniques of management, in relation to both the research done at any one site and the way that research was presented to the wider primatological literature and appropriated within it. To return to Carpenter's conditions, an undisturbed site where the observer could monitor the animals objectively without affecting their behavior, and thus produce a representative account of their activity, was an ideal impossible to achieve in practice. But compromises could be made, particularly by accounting for the particular conditions at a site and by placing them in comparative context. Field accounts had to include detailed accounts of the ecology of the site, report how much that environmental context was thought to have been affected by human influence either in the present or in the past, describe the attitude the primates showed toward human presence, and estimate how that presence affected their behavior. Fieldwork was to take account of this particular combination of conditions, unique to that specific site, while at the same time following the general principles of quantification and representative sampling that would let the results be judged in relation to the extant literature. In a sense, each field site represented a "natural experiment" where behavior could be recorded under certain conditions and compared with that at other sites where the factors considered relevant to understanding a particular behavior, or set of behaviors, would vary. In fact, variability became the dominating theme in understanding primate behavior. As more studies were carried out on new species and on old species in new locations, the provisional nature of the early conclusions regarding the nature of primate sociality, the relationship between the individual and the group, and the adaptive basis of primate behavior became evident to field researchers and opened up new problems to explore.

These questions and others relating to primatological fieldwork during this early period will be considered in the next chapter. But at least three key points emerging from this overview of field methodology are central to the discussions of the infanticide controversy in part 2 and to the understanding of the fieldworker's regress. First, the question of what was to constitute the "natural" primate group remained unanswered, especially what precise degree of human disturbance would merit rejecting the results of fieldwork. Second, the relation between laboratory and field research was still unclear: in practice the two spaces were rarely as directly opposed as they often were portrayed in idealized discussions of methodology, yet the standards of appropriate behavior for each still differed sharply. Finally, the duration of a field study, the number of hours of direct and personal observation of the animals that the researcher should achieve, and the decision on what type of behavior the project should focus on varied according to researcher and field site: none of these points could be standardized. In practice, the naturalness of the group and whether it had been watched long enough and under the right conditions were issues hammered out between authors, reviewers, and editors, since the answers would also depend on the species and the site. However, the absence of a consensus meant that all these issues *could* be raised as criticisms of a study—if the audience for a particular report felt it was necessary.

TWO

Studying Primate Societies, 1930–74

Introduction

As the last sections of chapter 1 demonstrated, "place" had become increasingly significant to the debates surrounding the search for field methodologies in primatology. Behaviors observed and work done at a field site in, say, South Africa, were clearly not going to be precisely recreated or replicated by researchers in a laboratory or even at a field site in Uganda or Nepal. Scientists were thus required to put enormous trust in the ability and willingness of individual researchers to carry out their observations appropriately and report them accurately.[1] In some cases this could be done through personal warranty, but as the nascent primatological community grew, the markers of reliability outlined by Carpenter and pursued by later writers—the detailed account of the ecology, history, and social setting of the field site, together with a reflexive reporting of and accounting for methodology and personality—were to become increasingly important. This was not only because the contextual information made assessing the trustworthiness of these reports possible, but also because their very existence in such a wealth of detail meant that by the early 1970s it was becoming feasible to treat the extant literature as the source of crucial comparative data and as part of a functional alternative to the laboratory notion of replication. As the previous chapter explained, carrying out experimental work at a field site was problematic: but if systematic and standardized reports of behavior from many sites existed, then it should still be possible to isolate and examine the variables underlying animal social structure by comparing their expression across sites. Success depended on carefully recording exactly those contextual details that Carpenter emphasized, which would—theoretically at least—tie each site firmly into a network of comparative relationships so individual observations could be assessed and interpreted against a much wider background.[2]

But as the previous overview of methodological questions showed, this position had not been reached easily or unanimously, especially as questions of similarity or variability became central to the discussion. Here again, the disparate origins of primatology and primatologists were important, as was the public reception of primatological information: while some researchers emphasized similarities of social structure both within and across species, others pointed to the steadily growing evidence of primate variability even between different groups of the same species. Not only social structure but the origins of social life were avidly debated as researchers tried to discover why most primates—unlike most other mammals *but* like humans—tended to be found in relatively permanent social groups that included members of both sexes and all ages. Again, the range of answers offered can be traced to the institutional origins of primatology as well as to the nature of the evidence that became available during this period. And finally, from the mid-1960s on, the theoretical understanding of the organization and origins of society and sociality were also being transformed as the concepts and perspectives that eventually became "sociobiology" coalesced. This was to have particular resonance for primatology, for reasons to do with its multidisciplinary and international origins, the importance attributed to social relationships in the understanding of individual primate behavior, and—most obviously—the perceived relation between the nonhuman primates and humanity itself.

This chapter will show how important this trilogy of interrelated issues—social origins, social variability, and sociobiology—was to the infanticide controversy. But throughout this period researchers still focused on the questions about primate society first raised by investigators such as Zuckerman and Carpenter. Discussions of the role of sex and aggression in society, the importance and function of social roles, the relation between the evolutionary history of the species and the behavior of the individual, continued to dominate, and the proposed answers were consistently treated as relevant to the origins of human society. But the nature of the proposed answers inevitably changed as more was learned about primate society. These changing conceptions of primate behavior also provided the empirical roots of the infanticide controversy. Within the topic of infanticide, key themes coincided: the place of sex in primate society, the role of sex and dominance status in maintaining and disrupting social bonds, the nature of the contacts between groups of animals within the same geographical area, the characters of the social role(s) played by different age-sex classes within the group, and the general relation between ecology and social structure. As chapter 3 will show, all these issues had parts to play in the explanations for infanticide that were initially offered. The fact that no consensus existed as to their nature or significance meant Hrdy's adaptive hypothesis was launched onto uneasy waters. That her account was also thoroughly implicated in the developing debates surrounding the place of sociobiology in explaining social life made it unsurprising that the study of infanticide became controversial. On the contrary, given the circumstances of its own origins, it would have been astounding had it not.

This chapter will consider these topics as they emerged in the reports of primate field studies from 1930 to 1974[3] and will show how these debates created a situation in which reports of infanticides caused by a sexually selected strategy for maximizing individual reproductive success could cause such contention. Many of the questions about the nature of primate behavior and society tended to focus first on why primates, unlike many other mammals, live in relatively permanent bisexual groups, often including more than one adult male. Three main areas of debate can be identified in this context: the role of sexuality, the development and use of the concepts of "social instincts" and "social roles," and the significance of group integrity. I will examine these questions and their development before moving on to consider the implications the growing awareness of the extent of primate social variability might have for understanding primate society—especially given the desire to generalize from nonhuman to human primates. Finally, this chapter will outline the tentative conclusions about the origin, structure, and maintenance of primate society that had been reached by significant sectors of the primatological community, particularly with regard to the relationship between adult males and infants, and will briefly relate them to the new sociobiological synthesis and the appropriate level at which selection was thought to act.

Sex and the Origins of a Bisexual Society

One of the more unusual aspects of primate society is that most primates are found in relatively permanent groups that consistently include adults of both sexes. That is to say, males do not seek out groups of females only during the breeding season in order to mate, as do other complex and highly social animals.[4] Instead, the sexes actively associate throughout the year. Many of the early explanations of primate behavior focused on this characteristic feature of primate social life, not least because such relatively permanent bisexual groups also characterize present, and presumably past, human societies. There was particular interest in the *number* of males in the group, since again, the presence of more than one adult male was typical for human associations and was thought to signal an increased social complexity that would render such groups particularly relevant in understanding the evolution of human society, as Eisenberg, Muckenhirn, and Rudran (1972) noted.[5]

One of the first and most influential accounts of the origins of primate society and the permanent association of the sexes was developed by Solly Zuckerman. Zuckerman, recall, had little time for most descriptions of primate social life, since in his opinion they consisted of conjecture and tautology: trained as an anatomist and rejecting appeals to any woolly concept of a "social instinct," he turned to animal physiology for what he considered objective evidence for social life. In his opinion this was the ultimate source of animal response and motivation, and it was in particular the *reproductive* physiology of the animals that was crucial to understanding behavior. The critical difference between primates and all other animals was that "monkeys and apes, like man, experience a smooth and uninterrupted sexual and reproductive life" (1981/1932, 51). While the "lower mammals" have times during the year in which it is physically impossible for females to conceive, primates can mate and produce offspring at *any* time of the year. Naturally, individual females would cycle through fertility, pregnancy, and lactation, but as one female passed out of sexual receptivity, another would enter it: hence the "main factor that determines social groupings in sub-human primates is sexual attraction" (31). Males are, in other words, constantly and consistently found in proximity to sexually receptive females: no other explanation for primate group living is necessary. This opinion was supported by Zuckerman's only adult attempt to watch wild baboons:[6] on a farm outside Grahamstown, South Africa, in May 1930, he shot and dissected twelve adult females and showed that all were at different stages of the reproductive cycle. This discovery did not just provide empirical confirmation for his belief that sex was central to society; it also demonstrated to his satisfaction the superiority of his methodology. From his perspective, "overt behaviour [should be treated] as the result or expression of physiological events" (1981/1932, xix). The cause of behavior must be sought within the internal environment of the baboon body: observation must be followed by physical demonstration. The ecological or social circumstances of the group were barely mentioned as influences on behavior, perhaps not surprisingly for a researcher more attached to the anatomy laboratory than the field site.[7]

Zuckerman's description of a sex-based society of free-living primates, originating in the anatomy of reproduction, had considerable authority, especially for researchers working with captive animals and for those who were particularly concerned to translate knowledge about primate behavior into an understanding of human evolution. Even Carpenter, the pioneer in the development of field research despite his commitment to studying behavior within its detailed ecological and social context, concurred at least in part with his conclusions, arguing that "the process of group integration through sexual behavior is repeatedly operative, establishing and reinforcing inter-sexual social bonds" (1964, 66 [1934]): it "would seem reasonable to suppose that the relative constancy of primate groups is importantly related during all seasons to the incidence of sexual behaviour" (1964, 353 [1942]). Later writers, like Chance and Mead,[8] cited both Zuckerman and Carpenter approvingly for their demonstration that "sexual attraction provides the fundamental bond uniting the individuals of primate societies" (1953, 415), a factor that for them represented a "key aspect [in the] ascent of man" (437). Marshall Sahlins, in his influential account of the relation between human and nonhuman primate social structure, argued explicitly that primate society was based on sex, since it was the development of the physiological capacity to mate at any season that "impelled the formation of year-round heterosexual groups among monkeys and apes" (1959, 56).[9] However, while Carpenter had agreed with Zuckerman's claim that sexual attraction between males and females was an extremely important source of primate sociality and mutual interdependence (1964, 375 [1952]), he did not by any means consider it the only fundamental element of primate social life. Later writers were to challenge the idea that sexual activity could play any integrative role at all in primate society.

In fact, as the number of systematic field-based studies of primate behavior increased, Zuckerman's contention that sexuality played a permanent and fundamental role in primate society began to receive direct and sustained criticism. Jolly's work with the lemurs of Madagascar had demonstrated that for these primates a breeding season *did* seem to exist (Jolly 1966b), as had Conaway and Koford's analysis of the breeding records of the Cayo Santiago macaques (1964) and Imanishi's account of the behavior of Japanese macaques (1960). Lancaster and Lee's (1965) review of the available field data on primate reproduction tentatively concluded that either a breeding season or a birth peak did in fact appear to exist for most species of monkey studied. In opposition to Zuckerman, they concluded that it was "clear that constant sexual attraction cannot be the basis for the persistent social groupings of primates" (1965, 513).[10] Indeed, for some researchers (Washburn and DeVore 1961, 1962), sexuality was not only *not* the prime source of primate social life but was a profoundly *disruptive* factor. Its operation had to be carefully mediated through social structures, such as dominance hierarchies, that prevented sexual activity from destroying group cohesion or males' capacity to form bonds directly with each other (Altmann 1959; Chance 1955, 1962; Imanishi 1960). Increasingly, explanations of the permanent societies of the nonhuman primates that drew on the operation of factors both internal and external to the group began to appear, accounts that treated sociality as subject to both stabilizing and disruptive pressures and considered not only the character of the various relationships that appeared to exist between members themselves, but also the way the group acted as a medium relating individuals to their wider environment.

Attraction, Dependence, and a "Social Instinct"

So important did the group seem for the individual that some observers resurrected the concept of a social "instinct" that Zuckerman had so roundly derided. Nissen, for example, described the way observers such as Kohler had identified a chimpanzee "need for companionship" (1951, 431) and suggested that the "need for social interactions is evidently one of the most powerful motivating forces among chimpanzees, and probably among other apes and most monkeys as well" (435). He refrained from speculating on whether this gregariousness represented an independent element in the emergence of sociality or was the result of an association between social living and the satisfaction of other primary needs, such as food and sex. In her observations of bonnet macaques, Nolte saw one female abandon her baby in her urgency to remain with the group as they fled from possible danger, and she speculated on the existence of a "social instinct to stick together" that might be powerful enough to override "mother instincts" (1955, 180). Altmann (1962a, 1962b) argued that while seasonal breeding had in fact been identified on Cayo Santiago, the animals were no less sociable outside that breeding season, and he suggested that researchers must look elsewhere for explanations of permanent primate groupings—perhaps in the findings of Mason that rhesus monkeys in stressed situations "were less likely to show responses indicative of emotional disturbance if they were accompanied by a familiar partner" (1968, 388). Jolly suggested that some kind of social instinct might be rooted in the long period of primate infant dependency: since primates retained many physical infantile characteristics, perhaps the "*original* cohesive force in primate social evolution would then be an infantile or juvenile attraction to others that was retained in the adult" (1966b, 162–63).[11]

Proving the existence of such an "instinct" would be hard, however, and other writers chose to seek empirical evidence for social cohesion in the relationships that structured primate society, continuing the tradition that Clarence Ray Carpenter had begun. As we have seen, Carpenter had given some support to Zuckerman's emphasis on the importance of sexuality to the primate group, but his own account of the basis of primate society was more nuanced. He not only acknowledged the importance of *both* external *and* internal factors in explaining the emergence of the relatively permanent group life of primates, but also considered nonsexual relationships of great significance to the balance of centripetal and centrifugal impulses that constituted the individual's experience. For example, in 1942 he described the maintenance of group integrity over time in the following terms:

> Males are dependent on females for sexual satisfaction and vice versa . . . females and young are mainly dependent on the adult males, although adult females and young may take an active part in group defence. Each group is dependent on the supremely dominant male, and to a lesser degree on his male associates to maintain freedom of movement within a limited territorial range and to guard against encroachment by other groups or by predatory animals. (1964, 355 [1942])

From this perspective, while sexuality remained a key element in accounting for the origin and maintenance of social life, it was only *one* element: females and immature animals were also drawn to adult males as the principal sources of group organization and as the group's primary defense. Later researchers elaborated on this model of primate social life as enmeshed in webs of mutual interreliance, examining the nature of the attractions and repulsions between different age-sex classes in primate society and the role these relationships played in maintaining the cohesion of the social group.

One fundamental point Carpenter identified in this early comparative study was that, though sex ratios appeared equal at birth, in most species mature females seemed to far outnumber males. Identifying this as the "socioeconomic sex ratio," he treated it as one of the key characteristics of primate society, an element that would remain relatively constant across different groups of the same species. And some later researchers came to treat this unequal distribution of males and females as one of the key points in accounting for primate social structure. Where did the extra males go? For some species they were known to exist in all-male bands outside the bisexual group, but not in all. How were males excluded from the group? In those species where more than one mature male was known to belong to a particular group, how were intermale relationships managed? Not only were these issues to became central to the infanticide debate, but the answers were also thought to be particularly relevant to understanding human evolution: species that showed a multimale social structure were also considered more complex, and hence more similar to the human experience. But researchers were still deeply unsure about why more than one male would be included in the group. What purpose did the extra males serve?

Security and Society

In fact, the function of the primate male was to be fundamental to one of the most influential models of primate social origins and primate social structure to emerge in the 1960s. This was the model developed by Sherwood Washburn and his graduate student Irven DeVore, based both on fieldwork done in Rhodesia and Kenya and on the theoretical framework brought to the field by Washburn in particular. Trained in anthropology, in the late 1930s Washburn had been a junior member of the Harvard Asiatic Primate Expedition, organized by Harold Coolidge and also including Carpenter. This experience confirmed him in the belief that understanding the nonhuman primates was fundamental to appreciating the nature of human society, and it also contributed to his formulation of what became known as the "new physical anthropology" (Washburn 1951).[12] This approach rejected the static, classificatory attitude that had typified previous anthropological accounts of human social relationships; instead, it privileged a dynamic, evolutionary model that placed human behavior firmly in the context of its primate heritage and subjected social relationships to exactly the same adaptive analysis that had traditionally been applied to anatomical structures. In this new anthropological worldview, behavior was just as much the product of evolution as was the body; concomitantly, human behavior could be understood only as part of a continuum of primate behavior. This close association between anthropology and primatology was to persist in the United States, and it meant that the pertinence of primatology to the understanding of human evolution was historically more salient for American primatologists than, for example, those based in Europe.

Washburn and DeVore had focused on baboon behavior and social structure for one fundamental reason. Many people thought baboons were the most likely source of inspiration for informed speculation about human origins. Baboon society and ecology seemed similar to what was thought to be characteristic of early humans: both species lived in large groups containing adults of both sexes and all ages; both were terrestrial, having abandoned the relative safety of the trees for the open savanna; and both were dietary opportunists. Baboons, like early humans, had to learn to manage the inevitable tensions in social life between individuals of different status, size, and motivations, and both would also need to orient both group and individual behavior to the occasionally conflicting demands of food acquisition and predator avoidance. In fact, Washburn and DeVore treated the group itself as the fundamental adaptation for individual survival: they argued that "seen against the background of evolution, it is clear that in the long run, only the social baboons have survived" (Washburn and DeVore 1961, 71). Additionally, this approach seemed to explain a number of social and biological elements of group life that had been noted by several observers.

In particular, it enabled them to account for Carpenter's "socioeconomic sex ratio" as a characteristic element of primate species society. In a given primate group, males were usually larger than females and outnumbered by them. Washburn and DeVore's model explained this by the way the needs of the individual animal were mediated through the group's relation to the wider ecosystem and with regard to the most successful distribution of the biomass of the species. Primate males, they argued, especially in species that were typically found in terrestrial habitats, had evolved to become significantly bigger than females, with far larger canine teeth and a concomitantly greater capacity for aggression. This was necessary, since males were needed in case of attack, as their description of troop movement showed:

> As the troop moves, the less dominant adult males and perhaps a large juvenile or two occupy the van. Females and more of the older juveniles follow, and in the centre of the troop are the females with infants, the young juveniles and the most dominant males. The back of the troop is a mirror image of its front, with the less dominant males at the rear. Thus, without any fixed or formal order, the arrangement of the troop is such that the females and young are protected at the centre. (Washburn and DeVore 1961, 63–64)[13]

The increased size and aggression of the males enabled them to defend the troop effectively: had females grown big enough to defend themselves, then far fewer of them could feed effectively on the same area of land. Through a division of labor in which many small females bore and reared young and fewer large males protected them, the group was able to maximize the number of infants produced and the survival chances of its members.[14]

However, when there were several large, aggressive males within a group, their relationships had to become far more elaborate and complex: these males had to find a way of managing their lives together without devastating violence. The development of dominance hierarchies seemed to be the solution, and studies of such hierarchies became a key element in primate reports from the early 1960s. As far as this concept's application to primate studies was concerned, in work done on captive animals in psychology laboratories, it regulated group organization, formalizing the competition between individuals for access to food and reproduction and minimizing the need for active aggression. The hierarchy determined the outcome of interactions: in normal circumstances there was no need for animals to fight for a desired object.[15] Males' aggressiveness, arising from their roles as the leaders and protectors of the group and the need to defend the group's integrity, itself had to be managed through structures that enabled them to live together without tearing the group apart.[16]

Survival, Predation, and the Male Primate

But what were these males protecting their groups from? Researchers in the 1960s identified three interrelated elements: threats to group integrity from others of the same species; threats to territorial integrity; and threats to individual survival from predatory animals. The debates focused again on the way the group structure mediated between the individual and the local environment, since these models of primate social structure assumed that primates formed groups both to defend against predators and to ensure regular access to food, water, and sleeping sites. The question whether primates exhibited any sense of territoriality was central, but it was hard to determine what "territoriality" might mean for them. Carpenter had concluded that for howler monkeys, "the possession of a territory is not a static but a dynamic adaptation . . . , that there is considerable overlapping [in] the territories of some groups, and that the ranges of some groups may be almost identical" (1964, 35 [1934]). Howler males, according to Carpenter, did "shift from one clan to another" (81), but any further contact between groups was highly limited. Instead, the eponymous howlers would vocalize at each other, using their howls to maintain physical and social space between groups. Altmann's later study of the same howler population agreed that the animals were organized "into quasi permanent, territorial, closed social groups" (1959, 321), and Sahlins's account of the evolution of primate society (1959) asserted that primate societies tended to be both semiclosed and willing to defend territory.

Not all observers agreed with these accounts, having found little evidence of conflict between groups. DeVore and Washburn's model dealt with this—as Carpenter had—by focusing on spacing mechanisms, which could obviate the need for aggression without rejecting the notion of territoriality outright. By 1963 DeVore felt able to conclude that "all monkeys studied to date live in organised groups whose membership is conservative and from which strangers are repelled. These groups occupy home ranges which may overlap extensively with those of neighbouring groups, but which contain core areas where neighbouring groups seldom penetrate" (311). Where physical aggression had not been observed between primate groups, social distance was maintained by visible or vocal displays, especially the latter for arboreal animals such as langurs, since visibility in the trees was limited. By 1966 Washburn argued that observers had systematically underestimated the importance of territory, partly because they tended to be more interested in interactions within a group, but also because spacing mechanisms were efficient enough to make it appear that there was no conflict: "The observer sees the results of avoidance, not the events causing it" (Washburn 2000/1966, 272).[17]

Similar problems of observation were raised concerning the predation threat—although it was even harder for observers to find unequivocal examples of males' repulsing predators. For example, Carpenter had noted evidence of predatory cats in the area where he studied howlers and had reported his field assistant's "dramatic and sincere" claim that a puma had successfully lured the monkeys close enough for an attack (1964, 84 [1932]). He explained his own failure to see a clear-cut case of predation by saying that these attempts were "probably frustrated more often than not by the clan males" (1964, 85 [1932]). Similarly, both Jay (1963a) and Rowell (1966) noted potential predators at their study sites but never saw violence. Hall argued that the "vigilance behavior" he observed in the male patas monkey was a response to predation threat, but again, she never saw an attack. Jolly's lemurs (1966b) gave an alarm call on sighting a hawk, but she never saw a hawk strike. Consistently however, the persistent presence of more than one male in the primate group and the occasionally pronounced sexual dimorphism between males and females were accounted for by predation threat. In 1932 Carpenter had argued that the "clan males" would defend infants: fifty years later Jolly asserted that males "play a special role in challenging predators, particularly if an infant is involved" (1972, 73).

As with territoriality, the lack of observed predation did not mean it was treated as less important in the evolution of primate society. Instead, primatologists were concerned to explain its absence. Some observers argued that human disturbance was reducing predation risk: deforestation and the increased exploitation of land for agriculture in a time of rapid human population growth meant that predatory species were swiftly becoming endangered. It was also possible that the simple presence of human observers was preventing "normal" predation on the groups they watched. So, for example, Washburn, Jay, and Lancaster (1965) stated outright that "attacks are seldom observed, because the presence of the human observer disturbs either the predator or the prey. . . . even today, most studies of free-ranging primates are made in areas where predators have been reduced or eliminated by man" (1542). Time of day also mattered: many if not most predatory species were nocturnal, hunting only when both the subjects of observation and their watchers were asleep. Rowell noted that most primate populations grew over the course of a field study, suggesting that the human observer was in some way protecting the subjects (1967, 224). Jay emphasized that all monkeys were still known to give alarm calls, presumably as an evolved response to past predator pressure (1965). In the light of later events, it is interesting that she went on to warn of the particular problems associated with observing rare, sudden, and violent events. As an example, she cites her own failure to observe predation among the langurs at Orcha and contrasts this to George Schaller's discovery of langur remains in the scat of the big cats he studied in a similar environment. Writing under the name Dolhinow, she argued that this example "points up how cautious the observer must be about assumptions of events that do not happen during a study. The chances of seeing a kill are so small that even with many hours of observations no kill was even suspected" (Dolhinow 1972, 196).

Despite the absence of recorded attempts at predation on primates, behavioral cues and circumstantial evidence were sufficient that primatologists continued to cite predation risk as a major impetus toward group living in primates.

Washburn and DeVore's "baboon model" encapsulated a number of elements common to primatological research in the 1960s. Zuckerman's anatomical approach had been rejected in favor of a perspective that like anthropology and ethology, two of primatology's most significant parent disciplines, focused on the individual within the natural environment. This did not mean laboratory work on captive primates was treated as irrelevant to the study of primates in the field: on the contrary, one of the most important concepts in fieldwork at this time was the dominance hierarchy, which derived from psychological studies of captive primates. But it did mean that the significance of the group loomed large in the individual's life: lone primates did not readily survive in complex and dangerous environments. Group life permitted the individual to live, and group life saw the beginnings of what was to become—for humans—a complex division of labor and an elaboration of social roles. Other than the maternal function, the most important role for the group was leader and protector—even where a threat had not been unequivocally demonstrated to exist. Washburn and DeVore's model was immensely influential—but it did not go unchallenged.

Critical Appraisals

Criticisms focused, first, on how far the model could be generalized to other primate groups, and second, on the adaptive role attributed to the males. This particular vision of troop structure was oriented overwhelmingly to the need to defend the troop from external danger; but not all primates (indeed, not all baboons) were found in ecological situations that required communal defense. For Hanuman langurs, for example, troop life was not necessarily a prerequisite for survival, as shown by the small number of males found outside the bisexual troop structure (Jay 1965, 249). And in her studies of forest baboons in Uganda Rowell found that the pattern of troop movement identified by Washburn and DeVore "was seen on occasion, but only when the cause of alarm was so slight that more confident adult males did not respond to something that set the juveniles running; a stronger stimulus produced precipitate flight, with the big males well to the front and the last animals usually the females carrying heavier babies" (1966, 362). Where real danger threatened, the adult males saved their own skins.

The idea that troop defense was the reason for pronounced sexual dimorphism was also challenged by Struhsaker, who argued that in the extremely dimorphic drills he had observed in Cameroon, there was "no evidence that adult males play any role in defence of the group" (1969, 97). Similarly, Hall criticized the notion that primate males acted as sentinels or leaders, first on the grounds that the terms themselves were anthropomorphic, and second, because there was little evidence to support such a contention. From his reading of the literature, "there are no grounds at present for assuming an 'invariable' vigilance by any individuals of a group" (1960, 280).

The adaptive role of aggression was also questioned, as was the significance of the dominance hierarchy, since the studies of free-living primates did not clearly support the findings of the psychological laboratories. For example, the relation between dominance and sex did not seem to be as clear-cut as the theory of the pecking order might suggest (Altmann 1962b; Hall 1963; Conaway and Koford 1964). For example, while the dominant male did seem to have most mating success, it was by no means clear that that this success descended linearly in the way the hierarchy implied.[18] In fact, when the data on the lives of nonhuman primates *other* than the baboons and macaques were considered, the importance of dominance in their lives and even the usefulness of the concept itself appeared debatable. For example, Emlen and Schaller's work on the mountain gorillas (Emlen 1960; Emlen and Schaller 1963; Schaller and Emlen 1964; Schaller 1965a, 1965d) emphasized that although a dominant male could be identified, the *lack* of clear dominance behaviors seemed to characterize these apes: peripheral males seemed to be accepted into the group without aggression, and dominance or aggression was rarely seen in relation to food or mates.[19] Other ape observers had equal difficulty establishing the leadership or social structure in the groups they observed (Goodall 1965, 454, 466; Reynolds and Reynolds 1965, 415–23). Jay's observations on langurs, a frequent point of comparison for the baboon and macaque data, acknowledged a dominance hierarchy for the males in the group, as well as a less well-defined hierarchy for the females, but stressed the lack of aggression in group life (1963c, 121).

In relation to the concept of dominance itself, some researchers tried to recover the situation by making a clearer distinction between the aspects of dominance as they might be expressed as part of group life. Aggression and dominance, while closely related, could be separated in that dominance hierarchies were meant to control the expression of aggression, so that species without clear dominance hierarchies either were less aggressive (gorillas) or, like lemurs, were aggressive for only a limited period of the year (Jolly 1966b).[20] However, Gartlan's (1968) review and critique of the concept of social dominance went further. He argued that its importance for the social lives of primates had been overestimated owing to the influence that studies of captive primates had exerted over the early field studies. He pointed out that that the application of the concept was often circular in practice,[21] that it was hard to establish linear hierarchies (pecking orders), and that the importance of hierarchies and levels of aggression was much more pronounced in captive animals, who had more leisure and less space than did free-living animals. Rather than treating it as a unitary characteristic, he suggested it might be better expressed as a role played by different animals at different times. To understand primate structure, one had to address the evident variability in levels of primate aggression and dominance behavior.

As the sixties progressed, however, more evidence of physical aggression between members of different primate groups began to appear in the literature, especially in relation to langurs. For example, Stuart Altmann's edited collection *Social Communication among Primates* (1967) included two accounts of langur behavior, and both discussed violent encounters between langur groups. The first, by Yukimaru Sugiyama, will be discussed in detail in the following chapter. The second, by Suzanne Ripley, described the territorial behavior of the langurs of Ceylon, arguing that the troops would seek out aggressive encounters even though they possessed the same spacing mechanisms that allowed other langurs to lead relatively relaxed lives. Similarly, Frank Poirier (1969) argued that Nilgiri langurs would violently defend against territorial incursions. Despite Alison Jolly's suggestion in 1972 that these examples of aggressive defense might result from human environmental pressure, and Bates's conclusion (1970) that the application of the concept of territoriality to primates was itself seriously flawed, by 1974 primate hostility to outsiders was well enough established for an edited collection of articles on primate aggression, territoriality, and xenophobia to be published (Holloway 1974). At the beginning of this period the aggressive defense of territory, whether physical or social, had been taken as given. Having received sustained criticism over the sixties from some of the most respected American primatologists, by the middle seventies it was buttressed by apparently solid empirical support from both field and laboratory sources.

But questions remained. What was the impulse behind this aggressive defense of the group's social and physical boundaries? Were males defending the group, or were they defending particular group members? In the cases described above, and in others, the males were usually described as concerned to defend the vulnerable members of the group—in particular infants and females with infants. Booth's account of the behavior of vervet monkeys in Kenya included the observation that while all adults reacted strongly to the presence of very young infants, the dominant male of the troop took the most active role in trying to protect them (1962, 484). In the case of Japanese macaques, Imanishi (1963/1957, 77) argued that the relationship between the central males and the group's infants was so intense as to justify describing it as paternal care.[22] In contrast, Jay's langurs showed little or no interest in infants. However, Jay emphasized that, like baboons, mothers with infants "travel . . . within a circle of the rest of the troop members, including the adult males, and in times of danger she and her infant are protected by the rest of the troop" (1962, 469). Even where the adult males were indifferent to infants, they were still considered vital for their protection. However, there were also examples of adult males' taking a rather different attitude toward infants. Southwick, Beg, and Siddiqi saw adult males frequently "attack infants or juveniles, particularly at feeding times. If an infant or juvenile got in the way of an adult male who was feeding, the adult would often attack it, picking it up, biting it, and throwing it to the ground" (1965, 153).

To sum up, by the early 1970s a variety of proposals had been put forward to explain the origins of primate society. In the intense debates surrounding this question, apparently unassailable explanations had been found, if not inadequate, then certainly limited in their application. The certainty with which Zuckerman had proclaimed the centrality of sexuality to primate society had been badly shaken by the evidence that several primate species experienced periods of group asexuality. The second "baboon model," that of Washburn and DeVore, had been found wanting in both its applicability to other species and its accuracy in the case of baboons. It seemed that for every primate group that behaved in a way that supported hypotheses concerning primate social origins and group structure, other animals behaved so as to refute them. Generalizations across taxa, and even from group to group within a species, seemed to become more precarious as more field-based reports of behavior were added to the literature.

The Unpredictable Variety of Being a Primate

Variability had been a problem from the very beginning of systematic field studies of primate behavior. In 1942 Carpenter had acknowledged that, in population structure there existed "marked variability from genus to genus among the non-human primates" (1964, 347), but he argued that it was still possible to establish an average size and range of variability for each genus and perhaps for each species. Ten years later he suggested that the social behavior of species within the same primate genus would be similar enough to enable "relatively limited sampling of representative species of each genera [to] provide a reasonably adequate framework of data for conceptualising the main features of the patterns and functions of the genera of the non-human primate societies" (1964, 365). As late as 1962, other researchers were simultaneously recognizing and downplaying the importance of primate variability. So, for example, Southwick suggested that while behavior might vary both "within and between species," it was still apparent "from many field studies that certain types of inter-group relationships are more or less characteristic of certain species" (1962, 436). Taking a slightly different approach, Jay argued that it was still possible to see "fundamental social patterns [which are] common to all monkeys, but various components of these patterns are emphasised differently in different species" (1962, 468). For this reason, from her perspective it was essential not to assume that the behavior of the well-studied baboons and rhesus monkeys could be taken as representative of all primates (1963b, 283).

The tendency to attempt to identify species-specific characteristics was one that primatology had come by honestly: the search for fixed action patterns—stereotypical or characteristic behavioral structures—was a key element in ethological research. But it was to prove very problematic in its new context. By the mid-1960s, a number of voices were warning that generalizing about primate behavior was dangerous. Bourlière (1962) cautioned that too few systematic studies had been undertaken to warrant generalization. In contrast to his earlier position, Southwick became concerned that "serious dangers may exist in generalising from one species to another, from one primate society to another, or from one habitat to another" (1963, 5) and emphasized that social behavior is variable and adaptable, not least because primates could learn and could pass on traditions to subsequent generations.[23] Miyadi (1964, 786) echoed this by stressing that individual primates were not uniform in their behavior—Why then should the aggregate behavior of primate troops not differ? The fears of these researchers can best be summed up by Plutchik, whose review of the literature found "limitations and inconsistencies which prevent generalisations and make comparisons between studies difficult" (1964, 67). That this was not, in his opinion, simply the product of methodological inconsistency is illustrated by his quoting Seward to illustrate that "among primates the only generalisation is that generalisation is impossible. . . . There is no single primate pattern" (1964, 69). But did this mean primatologists would have to confine themselves to individual descriptive studies? And if so, how were they to claim a place for themselves as scientists?

The contributors to DeVore's influential edited collection *Primate Behaviour: Field Studies of Monkeys and Apes* were aware of this dilemma. It is notable that they appear to be doing their best to acknowledge and at the same time to deemphasize the significance of variability—at least for the monkeys. For example, in the papers that compare the data on savanna baboons gathered by Hall in South Africa and Washburn and DeVore in East Africa with those on hamadryas baboons acquired by Kummer and

Kurt in Ethiopia, while the sharp contrast between hamadryas and savanna social organization is acknowledged, the emphasis is on the underlying commonality—the adaptability of the baboon to a wide range of ecological settings. Although the authors accept that major differences in baboon behavior may yet be found, they argue that at present the data seem to demonstrate that the behavior of baboons in East Africa and in South Africa is very similar (DeVore and Hall 1965; Hall and DeVore 1965, 110). Echoing this, Jay argues that the appearance of variation in langur behavior may well be a methodological illusion, since "additional field studies will undoubtedly demonstrate that the social behaviour of all langurs is basically very similar, and that variations among the different kinds of langurs are the result not so much of truly diverse forms of behaviour as of different degrees of emphasis on patterns behaviour common to most langurs" (1965, 247).

All langurs, regardless of subspecies, were thought to resemble each other in their behavior more closely than they did baboons or macaques. When one turns to the behavior of the apes, however, a rather different story emerges. The behavior of the two chimpanzee populations described by Goodall and by Reynolds and Reynolds clearly differs sharply, although they acknowledged that this might result from differences in methodology or the fact that Goodall's account of the Gombe chimps "applies specifically to a chimpanzee population in a rather atypical habitat" (Goodall 1965, 472). But it seemed clear that the behavior of chimps and gorillas contrasted sharply (Reynolds 1965), and Schaller[24] was deeply concerned to stress that the "behaviour of primates is highly adaptable and that generalisations based on a few observations in one area may be quite misleading" (1965a, 367). In fact, in his concluding section he went further and argued that "extreme caution must be exercised in making generalisations about one kind on the basis of what is known about another" (Schaller 1965b, 480). Especially dangerous here was the tendency to extrapolate to human behavior based on generalizations about primates.

Overwhelmingly, from the middle of the 1960s, accounts of primate society underscored two elements: the existence of variability and the persistence of order. Although animals belonging to the same species were seen to behave differently in different places, these deviations were not random; different environments elicited and accentuated different aspects of a recognizable underlying pattern of behavior. Perhaps the best account of this approach can be found in Washburn, Jay, and Lancaster's summative analysis of the state of primate field studies, published in *Science* in 1965. They suggested that the variety was the result of limited data: as more long-term studies were carried out on more representative samples of primate species, a clearer sense of order would emerge. The danger, for them, was that what "might easily have been described in a short-term study as a species-specific difference of considerable magnitude [may turn] out to be the result of seasonal and local variations in food source" (1965, 1542). For these authors, one of the most important discoveries made by the scientific study of nonhuman primate life, in contrast to the earlier "travelers' tales" was the existence of social cohesion and stability. As they argued,

> There could hardly be a greater contrast than that between the emerging picture of an orderly society, based heavily on affectionate or cooperative social actions and structured by stable dominance relationships, and the old notion of an unruly horde of monkeys dominated by a tyrant. The 19th century social evolutionists attributed less order to the societies of primitive man than is now known to exist in the societies of monkeys and apes living today. (1965, 1545)

The appearance of variability meant more studies were needed, but the root of that variability was not unpredictable: its explanation was to be found in the particular context of the local environment of a given primate group.

This turn to habitat and ecology as the underlying solution to the problem of primate variability was not novel, but it became more important from the 1960s on. It was the basis for one of the first attempts to systematize what was known about the behavior of the nonhuman primates.[25] Thelma Rowell stressed not just present-day conditions but the need to recognize that habitat changes over time. She was concerned that in the absence of adequate knowledge of the "effects of environmental differences on primate behaviour, we may also be misled about evolutionary mechanisms in the group" (1966, 363–64), and she was emphatic that "few field studies have considered the effect of different habitats on a species" (1967, 226). John Crook and Stephen Gartlan, two researchers based in the psychology department at the University of Bristol, took things a step further in 1966 when they published in *Nature* their attempt to produce "grades" of primate society based on use of habitat. This attempt took variability as standard, implicitly criticizing earlier accounts for concentrating "on similarities rather than differences between taxa" (1966, 1200) in order to generalize from studies of single species. Their focus was overwhelmingly on the relation between social structure and ecology, and on the way particular ecologies had produced particular social structures. They examined the range of known primate behaviors and arranged them into grades that not only corresponded with the territories the animals inhabited, but were also linked to a progressively more complex set of social structures. As diet became less specialized and as the animals moved farther from the trees, more specialized social roles were needed, especially with regard to the relationships of the increasing number of large, aggressive males in the group.[26] Not surprisingly, the emphasis on steadily increasing social complexity encouraged one to see the organization of primate societies as progressively developing toward the human climax.

Levels of Selective Pressure

Crook and Gartlan's socioecological model had taken Washburn and De-Vore's use of the environment to explain the evolution of primate society a step further. Both models spoke of the evolution of behavior but treated the "struggle for existence" as between the individual and the local ecology, mediated through the structure of the group. At the same time, however, a revolution was occurring in the theoretical conceptualization of animal behavior and the basis of animal social organization, a revolution that was to turn the spotlight from the struggle for existence to the struggle to reproduce. Although the formation of mutually supportive roles within a given group structure continued to interest primatologists and the problems of socioecology remained fundamental to understanding the expression of behavior within a given environment, ideas were abroad that would reassess the notion of evolutionary competition. They treated it not as a competition between the individual and the environment to survive and reproduce, but as one between genes for representation in the next generation. In several ways, this transformation would prove particularly difficult for primatologists—especially, perhaps, in the anthropologically oriented communities trained and influenced by Sherwood Washburn and his graduate students in the United States.[27]

While the "modern synthesis" of the thirties and forties had led to the resurgence of a more self-consciously Darwinian biology as field naturalists and experimental technicians combined their understanding of how evolution might occur in populations and individuals (Mayr and Provine 1980), before the 1960s there was still considerable latitude for questioning the *level* at which natural selection might be expected to operate. In particular, during the late fifties and early sixties, biologists debated "group selection." If evolution was concerned only with individual success and survival, it seemed hard to account for the many examples of animal "altruism," where creatures appeared willing to sacrifice not only their immediate interests but even their long-term survival to benefit others.[28] One of the most passionate proponents of group selection was V. C. Wynne-Edwards (1962, 1965), an ornithologist from Aberdeen, who produced the most rigorous account of how such a process might operate in relation to natural selection. At the core of his argument was the assertion that the social structure of animal groups acted to regulate population levels. In essence, he contended that all groups are limited in their physical expansion by the resources of the environment they inhabit: the amount of food and shelter available is always finite. He suggested that populations that had developed mechanisms for limiting population size, such as restricting breeding to certain classes of individual, would be able to use their local resources more effectively. Thus, while in the short term indiscriminate breeding might appear successful, ultimately the excess numbers would damage the local environment beyond repair and endanger group survival: in the long term, prudence would prevail. However, this kind of explanation explicitly required that animals should sacrifice their own interests for the good of the group, which not only tended to contradict some of the most fundamental tenets of Darwinian biology, but also failed to identify any obvious physiological mechanisms through which it might operate.

Wynne-Edwards's work was one of the last attempts during this period to justify group selection. By the early 1960s the threads that would eventually be woven into the new sociobiological synthesis had begun to appear in the literature, beginning with the work of William Hamilton, then a student at University College, London. Hamilton's work, though initially marginalized, attempted a thoroughgoing revision of the classic Darwinian concept of "fitness," arguing that simply considering an animal's own reproductive rate was insufficient to account for evolutionary change: instead, he introduced the idea of "inclusive fitness" or "kin selection," which combined an individual's reproductive success with that individual's effect on the reproductive success of its relatives. Published in the *Journal of Theoretical Biology* (Hamilton 1964a, 1964b), his mathematical account was eventually translated to a wider audience through George C. Williams's immensely influential *Adaptation and Natural Selection* (1966). This blisteringly direct evaluation of group selection also attacked the uncritical use of the concept of "adaptation," arguing that it needed to be applied with far greater precision—telling "just-so" stories was insufficient and inelegant. A few years later, in 1971, Robert Trivers took the idea of kin selection a step further with the concept of reciprocal altruism, providing for the first time a theoretical justification for apparently altruistic behavior between unrelated animals, followed by the elaboration of the ideas of "parental investment" (Trivers 1972) and "parent-offspring conflict" (1974). Together with the work of John Maynard Smith on game theory (Maynard Smith 1976), these ideas represented the foundations for the new sociobiological synthesis, a synthesis that concerned itself with what the "gene's-eye view" of evolution.

However, these theoretical debates were still ongoing during the period this chapter is concerned with, and they have a particular salience for the historical development of behavioral primatology. For example, while primatologists denied that support for group selection had persisted in primatology any longer than in other areas of behavioral biology, nevertheless it seemed that hostility to the idea of genetic or individual competitiveness was remembered as common, although a distinction was drawn between researchers who were genuinely committed to a perspective that focused on "the good of the group" and those who were simply slipshod in presenting their ideas. So, for example, Thelma Rowell was sharply critical of the suggestion that group selectionist thinking had dominated primatology in the years before Wilson's *Sociobiology,* arguing that it was

> simply not true. And we, it is important to say, we in Cambridge certainly understood that selection acted on individuals; we didn't say that selection acted on genes, and I would still dispute that, but it does act on individuals. Now, that isn't to say that people didn't occasionally write sloppily, in ways that looked like group selection, or even thought sloppily about group selection. But if you put any of us into a corner, we would have toed the line about individual selection.

Individual competition was acceptable: competition at the level of the gene was less so. During the early 1970s, however, it was harder to make such careful distinctions. For example, in an influential textbook, Alison Jolly was willing to accept that if group selection was "weakened to that of kin selection [Wynne-Edwards's] arguments should hold. Social primates live in kin groups and there surely is differential selection among troops of primates as well as among individuals" (Jolly 1972, 269–70). How to demonstrate this was unclear, but group living seemed so central to the primate experience that it was hard to accept that cooperative social relationships were not central to the evolution of group life.

Indeed, as this chapter has shown, the primatological literature during this period was dominated by influential models oriented toward understanding the evolution of the individual within the group, not toward the genetic makeup of the next generation. But the increasing enthusiasm with which the ideas and concepts of the new, explicitly sociobiological synthesis were adopted during the early 1970s meant there was a strong possibility of antagonism between, first, approaches that tended to treat the group as the fundamental unit of primate society; second, approaches that focused on individual actions and reactions; and finally, approaches concerned with the level of individual reproductive successes across a given population. Clearly these points of view are not mutually exclusive, and it is possible to distinguish at least two if not all three perspectives in the accounts of primate society given by different primatologists at different times during the forty-odd years this chapter covers. Eventually, however, the infanticide controversy in its early stages came to encapsulate the conflict between them (though it was by no means wholly defined by it).

Conclusion

By the beginning of the 1970s, the pendulum of primate field studies had clearly swung away from the internal, biological explanations of behavior summed up in Zuckerman's anatomical autopsies. Instead, primatologists, having acknowledged the extent of primate behavioral flexibility, had turned to external sources to explain it, taking into account both ecology's influence on social structure and the influence of learning on the behavior of individual primates. Although some felt rather too much weight was given to local ecologies in accounting for particular social structures,[29] a consensus seemed to be emerging, especially among North American primatologists, that one of the most significant discoveries of the initial period of primatological field research had been the been "the recognition and description of variability in the behavior of groups within species that live in different habitats" (Dolhinow 1972, 20). That is to say, different groups of the same species of primates might behave differently, but the source of that difference was to be found in the interaction between different ecologies and the basic tool kit of primate behavior.

Given the key role Dolhinow was to play in the infanticide controversy over the next decade or so, it will be useful to review her conclusions on the nature of nonhuman primate society. In an edited collection explicitly intended to update DeVore's earlier *Primate Behaviour*,[30] Dolhinow provided a systematic account of the state of primatological knowledge at that time, an account that accepted the central significance of primate variability but ultimately focused on the similarities across species and drew heavily on the anthropological perspective of Radcliffe-Brown. For her, the key elements in primate social life were stability and predictability. Physical aggression was rare, range and territory were rarely defended unless there was "some unusual situation such as overcrowding or rapid change in the habitat" (1972, 360), and fighting to the extent of wounding was "exceedingly uncommon in a natural habitat" (360). The elements that make up social behavior were unlikely to vary: differences were seen in the way these elements were expressed, and these differences could be related to the habitats where the different primate groups were found. At the heart of the primate group, holding the group together, were the different social roles to be found in that group, roles that were not specific to a single individual but comprised the expectations individuals might have of each other. In the interweavings of these roles could be found the basis of social order, ensuring predictability and allowing the group to function appropriately. Adult males were "normally the protector[s], police[men] and leader[s], whereas the female is the mother, nurturer and follower" (1972, 379). While the dominant male could in principle do as he pleased, "through learning and experience these abilities are channelled in ways that are beneficial for maintaining the group and protecting members. Selection has undoubtedly rapidly eliminated those rare animals whose temperaments were extremes of anti-social aggressiveness" (379). In other words, the lives of successful group-living animals are characterized by cooperation, not by aggressive competition.

However, Dolhinow also had to update her own chapter on langur behavior (previously published in DeVore's *Primate Behaviour*) to take account of Sugiyama's observations of langurs at Dharwar—observations that had "documented some exceptionally interesting differences in social behaviour" compared with her own (1972, 235). At Dharwar, it appeared, the group leader changed relatively frequently, and a new leader "may kill all the infants in the group" (1972, 366). Dolhinow suggested that further studies were needed, since it was "difficult to understand how this frequent change of male leadership and the killing of young permits the survival of the langur population over time or how these behaviours are related to the habitat of the group" (1972, 366). And yet strange as these observations were in relation to the prevailing primatological climate, it appeared that Sugiyama was not the first researcher to have noted the marked aggressiveness of langurs under certain conditions.

PART TWO

The Infanticide Debates

THREE

Infanticide's Infancy

Introduction

Short reports on the behavior of langur monkeys date from the beginning of the nineteenth century and are remarkable for two main reasons. First, in common with most reports of animal behavior before the twentieth century, they are characterized by the exuberant anthropomorphism that Yerkes and Zuckerman criticized so strongly, laden with assumptions about the relationships between the sexes and nonhuman animals' capacity for intentionality and careful planning. However, these accounts also reveal a series of sharp disagreements about langur social behavior that prefigure the controversies of the twentieth century. So, for example, writing in the *Proceedings of the Asiatic Society of Bengal* in 1884, the commentator T. H. Hughes was critical of an earlier writer who had described the way

> at a particular season of the year the great body of he-monkies, which had been leading a monastic life deep in the woods, sally forth to the plains and, mixing with the females, a desperate conflict ensues for the favours of the fair lady pugs. This continues for several days, at the end of which time one male more valorous or strong than the rest will be found in possession of the flock, his discomfited fellows remaining at a short distance from the scene of their defeat . . . [T]he female monkeys [then deliver] up their half grown male offspring to the care of the former, who troop away to the jungle reinforced by the hopeful juniors, who the next season return with their foster fathers to take part in the contests. (147–48)

For Hughes this interpretation was incorrect: in his experience, although langurs did fight, it was "for reasons other than the luxury of a train of wives" (148). He believed these conflicts were initiated by the "incursion of a stronger troop into the domain of a weaker one" (150): in other words, the animals were contesting territory, rather than access to females. Hughes was not to have the last word, however. Just a few years later, in 1901, J. F. G. described how, while hunting, he had come across a group of "lungoors" that were greatly agitated by "the return to the herd of all the young males, which are yearly driven away by the principal male monkey, the strongest and the biggest, and, therefore, the sovereign, generally called 'the rajah.' I had heard from the natives how this happened every year, and that the rajah never kept undisputed possession of his harem for more than one year, and it was my good fortune to see how he was driven away or killed" (J. F. G. 1901, 149). Although these occasional observers of animal behavior differed sharply in their interpretations of the events they documented, it seems unequivocal that intense aggression between langurs had been witnessed and recorded several times by the early twentieth century.

Sixty years later, when the modern period of field-based primatological research began, these fragmentary accounts of the aggressive nature of langur life were dismissed along with most of the other "travelers' tales" about the behavior and habits of apes and monkeys. In particular, Phyllis Jay singled them out for condemnation in her introduction to her 1963 doctoral thesis, arguing that

> the incidents described in these accounts were anecdotal, often bizarre, certainly not typical behaviour. Constant reference was made to harems and to male overlords fighting to maintain absolute supremacy while forcing all other adult males from the troop. In several accounts, after the anthropomorphic interpretation is taken into consideration, the description of events is plausible, but the behaviour of a troop must be observed for months before the frequency and importance of events can be understood and evaluated (1963a, 8).

For Jay, the early observers had made a simple error when they focused on the aggressive behavior, one that could easily be explained. Langurs are arboreal creatures, and making systematic observations of them in the forest is very difficult. Thus it is natural that only the most spectacular and dramatic behaviors would have been seen and thought worthy of note by the hunters and travelers of the nineteenth century. When langurs were behaving "normally" or "typically"—that is, feeding, grooming, resting—they would probably remain almost invisible to the casual observer. Certainly, Jay's observations suggested an entirely different picture of langur life.[1]

This chapter will examine the field studies of langur behavior that took place from the late fifties till the early seventies, focusing on the reports made by Phyllis Jay and Yukimaru Sugiyama. By 1974 the results of at least eight field projects had been published, but these two remain especially significant. First, they represent the first extensive investigations of langur behavior in the wild undertaken by North American and Japanese scholars; second, not only did Jay have a central role in the later development of the infanticide controversy, but Sugiyama was the first researcher to connect infanticide with sexual access to females, although he was not the first to propose the sexual selection hypothesis in print. In this chapter I will outline the field projects associated with these researchers and put them in the context of the later field investigations and the debates that developed about the structure and function of langur society. In particular, I will examine the strategies researchers used to report their discoveries: remember, at this point the field study of primate behavior was in its infancy, and studies were—by the standards of the twenty-first century—extremely short. As a result, researchers who wished to describe individual development or the nature of social organization were drawing on very limited data, in a context where agreement on appropriate standards of field methodology had yet to be reached. The practical and rhetorical techniques they developed to manage this unstable situation were to take center stage as the infanticide controversy progressed.

Jay's Observations: A Very Peaceful Primate (1958–63)

Phyllis Jay (later Phyllis Dolhinow), a student of Sherwood Washburn's at Chicago,[2] inaugurated one of the earliest of the modern field studies of primates when she began her work on langurs in India in 1958. Her observations were concentrated at two main sites: Orcha, a village in the state of Madhya Pradesh (now Chhattisgarh State), and Kaukori, a village near Lucknow in the state of Uttar Pradesh. The two areas differed radically in their ecology and the way local people used the land. Whereas Orcha was a forested area, with minimal human cultivation and a low population density, at Kaukori almost all the available land was cultivated, and the communities of both human and nonhuman primates were correspondingly much more crowded. These contrasts meant there was a sharp difference in the nature and quality of the observations it was possible to obtain from the sites, not least because langurs tended to flee instantly on sighting humans in Orcha's forests, whereas at Kaukori they were accustomed to a constant human presence and took flight only when chased or threatened. After her initial surveys, Jay chose four groups for intensive observation, three at Orcha and one in Kaukori.[3] The sharp ecological differences between these two sites provided a "natural laboratory for the investigation of variables in behaviour which occur in nature" (1963a, 10). By comparing the behavior of the langurs at the two sites, she hoped to identify the basic patterns of langur social development and also to begin accounting for any differences by relating them to the environmental conditions characteristic of the sites.

However, while the different site ecologies made it possible to develop a fuller understanding of the structure and functions of a langur group and the way this was sustained by the interactions of the animals integrated within it, for Jay there "appeared to be no important alterations in basic patterns of social behaviour between the two environments" (1965, 199).[4] Rather than comparing and contrasting langur behavior in relation to ecological variables, the overwhelming focus of Jay's early reports (1962, 1963a, 1963b, 1963c, 1965) was on the way the socialization and early experiences of each individual langur produced a troop characterized by relaxed and harmonious relations between its members. In sharp contrast to the aggression and violence seen in captive macaques or the clear dominance hierarchies and intensive agonism observed in free-living baboons, here she saw a new primate pattern, possibly rooted in these creatures' being largely arboreal, easily capable of escaping predators by taking to the trees. Unlike baboons (Washburn and DeVore 1961), for langurs the predation pressure was considerably relaxed. Hence not only was it possible (although rare) for some males to live as individuals outside the troops, but langur males did not exhibit the physical adaptations for group defense (extreme sexual dimorphism and so on) that characterized the ground-living baboons and macaques. Jay believed this had profound consequences for group organization and for the relationship between adults and infants within the group.

Not only were Jay's langurs very different from the other monkey species that had been observed by then, what she saw also contrasted sharply with the previous reports of langur social life. In the first place, all the groups she observed were multimale. Unlike those in the earlier anecdotal accounts, which discussed the aggressive maintenance of a one-male social system, Jay's langurs lived in permanently bisexual groups that included more than one adult male. Like all other observed primates, these multimale troops did appear to be organized according to a dominance hierarchy, but unlike most other primate species, this male hierarchy seemed to influence social life in an extremely understated way. Her reports emphasized that it was "established and maintained with a minimum of aggressive fighting" and that most dominance interactions were "accomplished in a nonviolent manner by the use of very subtle forms of threat" (1963c, 121). In fact, these threats were so subtle that it took her a considerable time to recognize them. This was very different from the spectacularly aggressive interactions of terrestrial primates, such as baboons, as were the males' attitudes toward the infants of a group. Whereas adult male baboons were fascinated by infants and seemed to act protectively toward them, adult male langurs showed "no interest in the newborn or small infant langur" (1962, 470). In fact, the "loudest squeals of a newborn do not draw the attention of an adult male" (1963b, 292).

Female langurs also differed from their baboon cousins. Like baboons, adult and subadult females were intensely interested in infants—but unlike most other known primates, a mother would allow the other troop females to handle her infant very shortly after birth (1963c, 117), retrieving it only when the infant seemed distressed. A female dominance hierarchy also existed, but it appeared to fluctuate considerably according to the reproductive states of the females involved, with status increasing during estrus and the last weeks of pregnancy. A female with a newborn infant avoided all forms of aggressive interaction, so much so that her "dominance interactions with other adults [were] suspended temporarily" (1962, 470), and at the first sign of trouble she would swiftly withdraw. In fact, Jay observed violent female aggression in only one context—where an adult male had accidentally frightened an infant. In that situation the "mother instantly threatens, chases and often slaps the male" (1963b, 292), again in sharp contrast to the relations and reactions between males and females in baboon troops (1963c, 121). Finally, the only other times Jay observed aggressive interactions between langurs were the rare occasions when males from outside the bisexual troop tried to gain entrance. Jay noted that the "fighting that occurred during these contacts was the most severe [she had] ever observed among langurs" (1965, 206–7).

With these few exceptions, Jay's langurs were seen to live peaceful and serene lives, where each animal had a clearly defined role within the troop structure. The notion of the "troop," as distinct from the individual monkey, is fundamental to understanding Jay's approach to langur social life. In these early papers she devoted much time to analyzing the socialization process that produced these relaxed, integrated adults,[5] and the structure of her PhD dissertation is modeled on the development of an individual langur,[6] weaving together her observations of different age-sex classes to provide a continuous narrative that charts the growth of an infant through juvenility to maturity to senility. In the final chapter of this dissertation, however, she turned to the underlying basis of langur social structure. Drawing directly on the work of the anthropologist Radcliffe-Brown (Jay 1963a, 224), she argued that a langur troop must be understood as a structure within which individuals play defined roles[7] that exist in continuing relationships with each other through time. While the performance of a role may vary depending on the individual involved, the existence of such roles is fundamental to social structure, since it is only through them that individuals' expectations of each other are met and understood—at least under "normal" social conditions.

In fact, her treatment of "normal" social behavior is itself interesting in this context, as are her reasons for taking this approach. In her conclusion, she argued that "normal" social behavior could be seen only about 25 percent of the time. For the rest, normal behavior was disrupted by troop disequilibriums such as birth (where the newborn becomes the center of attention for females), estrus (where females will remain with males and ignore other age-sex groups), and dominance fighting (where males are avoided by all except subadult males). These categories were accompanied by diagrams describing and gauging the intensity of the social bonds between age-sex groups in these contexts—in all cases, explicitly compared with the periods of "normal" behavior. She argued that while this attempt to treat the "social life of a langur troop [as] a functionally-integrated structure which displays an over-all continuity through time" (1963a, 239) was very different from the way zoologists and psychologists looked at animal behavior, it was nonetheless necessary. Particularly when dealing with primates, the categories, patterns, and dyadic interactions other observers noted had to be placed firmly within the context of an integrated social structure—especially if the ultimate aim was to compare the behavior of human and nonhuman primates. Notable here was the way the social structure of a langur group was treated as quasi-independent of the animals themselves, with individual behavior discussed as significant only insofar as it demonstrated effective socialization into "normal" social roles that ensured the smooth functioning of troop social structure. As Jay pointed out, this is a profoundly anthropological, rather than zoological, approach, and as such it represents another demonstration of the equivocal place that nonhuman primates occupy within cultures that insist on a clear demarcation between "human" and "animal."

Perhaps the clearest statement of Jay's position on the basis of langur behavior is in a paper she published two years after submitting her dissertation, where she argues that the "context for langur social life is a stable, well-organised group composed of monkeys of all ages and both sexes. Such a group is the basic unit of the species. . . . Individual members assume roles and activities that assure group cohesion and pacific intragroup relations. Langur groups can be characterised as peaceful and relaxed; their members are seldom aggressive and serious fighting is rare" (1965, 216).

Primate variability was clear—using roughly the same methods and from the same troop-oriented perspective, studies of baboons and langurs had revealed unmistakable differences in the role and function of dominance, aggression, and group living. Baboons lived together because ecological constraints meant they had to in order to survive. Langurs lived together because of the social bonds between individuals and because the troop was the context where the young could mature and learn to act as adults. The idea that there was, or could be, one single primate pattern for group living and social structure had been shown to be an oversimplification. However, just how variable langurs were would shortly become clear, as the results of another prolonged study of langur life began to appear in print.

The Japan-India Joint Project in Primates: Dispatches from the Front (1961–67)

This study was rather different from Jay's, not just in orientation and operation, but also in its results. As chapters 1 and 2 have demonstrated, Japanese primatologists had made major contributions to the development of the discipline in the postwar period, and in 1957 the Japanese Monkey Centre was the first institution to produce a journal (*Primates*) devoted solely to the behavior, anatomy, and evolution of the nonhuman primates.[8] The cultural context for primate observations in Japan, however, was very different from that in the West, inevitably influencing not just the topics researchers chose to study, but also the way they went about it. In particular, Japanese researchers during this period showed a greater willingness to intervene in the lives of the primates they watched, perhaps because Japanese culture attached less importance to a clear boundary dividing humans and all other animals.[9]

In any case, in 1961, two years after Jay had completed her initial research, the Japan-India Joint Project in Primates Investigation was established, and under the directorship of Denzaburo Miyadi this group of researchers[10] began to investigate the behavior of Hanuman langurs and bonnet macaques in and around the town of Dharwar in Mysore State (now Karnataka), southern India. This research was conducted over a two-year period, from April 1961 until April 1963,[11] concentrating on a stretch of road connecting Dharwar with Haliyal, a town about thirty kilometers to the west. The study of the langurs of the area began with a roadside survey conducted by Sugiyama, who used a small motorcycle to ride up and down the road, stopping whenever he saw monkeys and noting the location, numbers, and the age-sex classes of the animals for future identification. Most observations at this stage were made with binoculars, since despite the langurs' indifference to traffic on the road, animals in the forest tended to retreat swiftly from direct and prolonged observation.[12] This repeated survey indicated that forty-four langur groups could be reliably found along this road, and by the end of September, not only had the groups been identified but their home ranges had been mapped.

Immediately apparent from this initial survey was that, like Jay's langurs, groups could be divided into bisexual troops and males that lived outside these troops, usually in male-only groups.[13] However, strikingly unlike Jay's langurs, these bisexual troops were usually characterized by the single-male structure described in the anecdotal accounts discussed at the beginning of this chapter, and as a result, there were many more males living outside the troops than had been seen in northern India. Almost three-quarters of all bisexual troops (twenty-eight out of thirty-eight) contained just one adult male, and half of those remaining had just one "large full adult and the others were all young adults" (Sugiyama 1964, 17). Further short surveys around Raipur in central India found that almost half the troops there also followed the one-male pattern, and Sugiyama concluded "that the principle of the social organisation of langurs is one and the same in both the districts, Raipur and Dharwar" (Sugiyama 1964, 27). The problem at this point was to discover how and why this one-male pattern was established and maintained, since it seemed to be a significant element in the structure of langur society. However, by the end of the roadside survey in September 1961, the researchers had also seen a number of unusual events among the langurs—a series of violent contests between the adult males of bisexual troops and stranger males, contests with consequences ranging from the replacement of the leader male to the division of the troop (Sugiyama 1964, 27–34). The potential relation between this one-male pattern and these striking periodic upheavals was to occupy the Japan-India Joint Project—and Sugiyama in particular—for the rest of the study.

To investigate the relation between social change and the maintenance of the one-male troop structure, seven troops were selected for intensive study in the second stage of the survey (Sugiyama, Yoshiba, and Parthasarathy 1965, 74). This stage involved both observing and manipulating social structure, in the hope of discovering not just the langurs' social composition and home range, but also the nature of their social activities, the exact role the leader male played in a bisexual group, the place of sexual activity, its timing and intensity in group life, and the nature of the relationships between the one-male troops and the male groups that seemed to range within their territories.[14] Crucial to the investigation of the behavior of Japanese macaques had been habituating the groups and identifying individuals, much assisted by provisioning the animals. For the langurs provisioning proved impossible, but individual identification and habituation succeeded nonetheless. As Sugiyama and his colleagues pointed out, four of the troops studied had become "so habituated to our observation during the first survey that the existence of the observers did not disturb their daily activities, so long as the observers were 10 metres distant from them" (Sugiyama, Yoshiba, and Parthasarathy 1965, 77). Jay had also offered food to her langurs—although more in the spirit of monitoring their reactions to novelty than to assist in habituation—but the Japanese researchers took intervention a step further by physically removing animals from the troop so as to monitor the others' reaction. Be that as it may, many of their results as they pertained to individual behavior were similar to the ones Jay recorded. But the greatest difference between the studies was in the nature and frequency of social change, changes that seriously disrupted troop composition and constitution—disruptions never seen or recorded for Jay's stable, closed social groups.

Sex and the Single Male: Structuring Social Revolutions

The nature and meaning of social change for the langurs of Dharwar is directly discussed in three of the Japan-India Joint Project's publications (Sugiyama 1964, 1965b, 1966). I will examine them in turn, since each demonstrated a different strategy for studying and accounting for social change, strategies that were later to receive intense attention from the participants in the infanticide controversy. In the first case, Sugiyama presented an "ideal type" of social change, constructed by unifying a number of fragmentary observations made of the thirty-second, thirty-fourth, fortieth, and thirty-ninth troops from June till September 1961 as part of the initial roadside survey. Second, he offered a narrative-based account of naturally occurring social change in the thirtieth troop in the first two weeks of June 1962. Third, Sugiyama described a social change that the observers experimentally induced by removing the leader male from the second troop on June 20, 1961. Like Jay's reports, these accounts focused on the impact these changes had on the structure of the troop, not on repercussions for the individuals within it; and one must remember that at this stage all attention was on the causes and consequences of social change—*not* on any consequences that social change might have for the troop's infants.

Sugiyama's 1964 publication is the first public report of the activities of the Joint Project. It began by describing the research project, the Dharwar area, and basic langur ecology and concluded by describing the structure of social change:

> The process of the social change may be supposed as follows: (1) the existence of a normal type of one male troop, (2) the attack and intrusion of a male group on it, (3) a severe fight between the males, (4) the desertion of the leader male from the troop in consequence of the fight, (5) the occupation of the troop by the strongest male, and the separation of the male group from the troop, namely, the birth of a new one male troop, (6) the reorganisation of the new troop by the leader male and the stability of the troop organisation. (1964, 35)

Note that this was a reconstruction of partially observed events. In no troop were all of these stages seen, but each stage was observed—in no particular order—in one or another of the troops contacted during the roadside survey in the summer of 1961. Sugiyama had sorted chaotic and fragmentary observations into a coherent, logical narrative, a narrative that could also account for the suspicion and hostility with which troop males regard male-only groups. As Sugiyama pointed out, since it was only "by attacking and occupying a troop [that] males in a male group can be a member and leader of the troop and get a firm [social and sexual] status. . . . [it] is natural that male monkeys should fight severely with each other to get the leader status of a troop" (1964, 36). There was no mention here of infant biting, infant killing, or infanticide. At this stage the model of social change Sugiyama presented was intended, first, to serve as a means of understanding why the one-male troop pattern was so common at Dharwar, and second, as a key to langur social organization. However, the possibility that social change may be dangerous to infants was hinted in the third Japan-India publication, which was devoted to individual behavioral development and within-troop social relationships (Sugiyama 1965a)[15] and dealt with social change only in passing. It concentrated on the second troop and described the life of a langur from birth through to adulthood—but since "all the infants in the 2nd troop died in an incident when the eldest infant became seven months old" (Sugiyama 1965a, 223), the descriptions of late infancy and juvenility were taken from the observations of the thirtieth troop. Sugiyama's audience was left in the dark as to the nature of the "incident" that eradicated the troop's infants.

In the same year (1965), Sugiyama also published the only complete account of "natural" social change that was seen to occur at Dharwar, which took place in the thirtieth troop between May 31 and June 14, 1962. In his introduction, Sugiyama reminded his audience that langurs at Dharwar are found either in one-male troops or in all-male groups and argued that once the process through which this one-male pattern was established and maintained is resolved, then the fundamental basis of langur social structure will be understood (1965b, 382). Unlike the artificial imposition of narrative on the observations that constituted the first description of social change, this article was structured chronologically, describing the day-by-day, hour-by-hour, and minute-by-minute events that disrupted the social order of the group over that fortnight.[16] At 2:00 p.m. on May 31, some of the females from the thirtieth troop were seen in the company of the male group. Over the next few days, the other males were able to drive the troop male (Z) away from the females, and by the fourth day after the initial contact, one of the new males (L) was beginning to act as the leader male of the thirtieth troop.[17] By day seven some females had disappeared, along with an infant, and another infant "had a cut 10cm long and 2cm deep on the left buttock. It was so serious a wound that he could not walk for himself" (1965b, 391). L continued to attack the mother of this infant, and Sugiyama assumed that "from the concentrative attacks of L on A and the severe resistance of A, that the cut on AI must have been given by L. Supposedly, Bi must also have been bitten to death by L. . . . But the dead body of Bi could not be found" (392). Even after the males had stopped fighting, L continued to attack the two remaining male infants and one of the female juveniles. All three were thought to be "very active and in good health" until they suddenly disappeared, never to be seen again (395–97). By August 1962, of the fourteen juveniles and infants that had previously resided in the thirtieth troop (six juvenile males, three juvenile females, three infant males, and two infant females), only two female juveniles remained. The six male juveniles had left the troop, and all five infants, along with the third female juvenile, had disappeared.

A week after the situation in the thirtieth troop had stabilized, the leader male of the second troop was captured and removed (Sugiyama 1966, 44). The researchers justified this direct intervention because the "observation of social change is not always possible in natural conditions" (1966, 42) and because understanding social change was fundamental to understanding langur society. Again, events were described chronologically, with excerpts from field diaries included, sometimes in the present tense, to illustrate particular passages directly. For the first week or so, nothing seemed to happen. The troop continued to forage and to roam as usual, but on the occasions where the troop met the other troops in the area—especially the fourth and the third troops—it appeared that the "leader male of each troop searched for . . . the leader of the second troop" (49). No male group inhabited the area, and at the time of the old leader's removal, the troop contained nine adult females, five juvenile females, two female infants, and two male infants. However, when the observers found the troop on the morning of June 28, a female infant was missing and a male infant was seriously injured. At this point the article begins to read like a murder mystery, as Sugiyama ponders the possible causes of the wounding and disappearance, concluding that "from the conditions of the wounds that *KI* must have been bitten by a strong male langur with his canines towards evening on June 27. Why did *Ui* disappear from the troop? It was supposed that *Ui* might have been injured like *KI* or killed" (1966, 49).

Predators had been excluded as culprits, and so "the highest possibility and suspicion was given to an adult male langur" (1966, 50).[18] But which of the males was most likely to be responsible? The only ones in the vicinity were the leader males of the fourth and the third troops. By the next day, two infants had disappeared and another was seriously injured.[19] Then, when the fourth troop met the remnants of the second, the leader male was seen to attack the mother of the injured infant.[20] By the next day, that infant had also disappeared. No other troop member had been injured. Over the next month, the leader male copulated with most of the females of the second troop but was unable to unify them with the fourth. Instead, by January 1963 they were to be found with the third troop, and four months later the erstwhile second troop females had given birth to infants fathered by the fourth troop's leader male. None of these infants were attacked by the third troop's leader.

The crucial observations at Dharwar had all been made by the early months of 1963, but their details were not widely circulated in the West until rather later in the decade. But what, essentially, had been seen at this southern Indian site? First, it was evident that not all langurs lived the peaceful lives Jay had described: clearly, there were circumstances in which langur society was liable to undergo considerable upheaval, and very aggressive interactions both between males and between males and females appeared characteristic of such periods. Second, despite the high levels of violence and though attacks were recorded on individuals that later disappeared, neither infants nor juveniles were *ever* seen to be killed, nor were their bodies found.[21] Third, at least one of the accounts of social change had been artificially produced by removing the troop's leader male, and the "standard account of langur social change" was produced by the combining fragmentary observations into a coherent narrative. Fourth, it was clear that infanticide—or infant biting, as Sugiyama called it—was peripheral to the interests of the Japanese-Indian research team, who were fundamentally intrigued by how the one-male social structure emerged and was maintained.

Discussions and Debates: The West and the Rest (1967)

These interests influenced the response that the nascent primatological community made to Sugiyama's reports. Since the circulation of *Primates* was limited in the West at this point, these accounts of infant biting were not widely known outside Japan until they were summarized and published in Sugiyama's contribution to Stuart Altmann's 1967 collection, *Social Communication among Primates.* Overwhelmingly, this paper concentrated on describing and diagramming the processes of violent social change in the troops of Dharwar.[22] His conclusion that a troop was likely to be "seized by social changes every three years, on the average [and that] these replacements of the leader male lead to the expulsion or death of all males, including juveniles and infants, from the original troop" (1967, 232) provoked a sharp response from the psychologist J. M. Warren, who had been invited to comment on the paper. While he did not doubt the accuracy of Sugiyama's reports, he emphasized that "the periodic liquidation of the baby langurs [is] potentially dysgenic in its consequences" (1967, 257) and that it remained to be shown whether this behavior was in any sense adaptive. However, this is the strongest, and almost the only, criticism of Sugiyama's accounts. His descriptions of infant biting and the circumstances in which it occurs are clearly mirrored in those produced by Hrdy and others nearly a decade later, but they did *not* provoke the controversy that Hrdy's papers produced. The reason seems clear. Sugiyama's intent was to explore the reasons a particular langur social structure might exist. He was concerned with infanticide, or rather, infant biting, only as a peripheral circumstance associated with rapid social change. Although Sugiyama was later cited as originating the sexual selection hypothesis as the explanation for primate infanticide, it is not at all clear that sexual selection played any part in his thinking at this stage.

This is apparent from the explanations for infant biting offered in these early papers. It was not until the fourth report of the Japan-India Joint

Project—the description of social change in the thirtieth troop—that the subject was broached, and then *only* in relation to the contribution it might make to maintaining the single-male troop structure. Having concluded that periodic attacks by the male group were the major mechanism for managing this,[23] Sugiyama considered the possible reasons for these male attacks. He decided that male groups attacked bisexual troops in order to "get females to satisfy their sexual urges and to have a living place abundant with food and safe from the enemy" (1965b, 408). In this context, for Sugiyama, infant biting was primarily a means through which the male can integrate the new troop. The desire for sex means that, having achieved residence in a troop, he

> must exclude all the nuisances, except necessary females [which includes juvenile females that will shortly be old enough for mating], out of the troop after he captures it, and he must integrate the females under his sway. . . . After capturing the troop, the new leader attacks and bites infants. He may treat them as the objects to be excluded from the troop like the other males. . . . Or he may need to attack and exclude them from the troop so as to make the strong social bond with the females. From these motives and others, the new leader may have bitten the infants as a kind of demonstration to strengthen the ties between him and the females and to integrate the females under his sway. (1965b, 414)[24]

In this interpretation, violence was not just necessary to oust the old male but was a key part of "making new and strong social relationships between the offensive adult male and the defensive females of the troop" (Sugiyama 1966, 68). This was why, after the experimental exclusion of the second troop's male, the leader of the fourth troop showed violent aggression even in the other male's absence. In short, males attack bisexual troops to satisfy their sexual urges, and the "severe fighting among males, change of troop leader, or a severe attack with killing of infants [will] activate the sexual activity of females" (Altmann 1967, 233). Sugiyama did note that without the eradication of the infants, the male "could not have satisfied his sexual urges" (1965b, 413), and that the "loss of the infant has the effect of advancing the estrus of the female" (Altmann 1967, 233); but while this sounds similar to the later accounts of infanticide, it was *not* an account of sexually selected infanticide. It related purely to the male's immediate motives and did not refer to his ultimate reproductive success. During this period Sugiyama was less concerned with infant biting itself and far more interested in the impact it might have on the sociosexual relationships between adult langurs and the reintegration of the one-male troop structure.[25]

In the same year, however, another publication appeared in *Laboratory Primate Newsletter*, linking infanticide and sexual selection in print for the first time. This was a short account of an infanticide that had been accidentally provoked—and filmed—in a psychology laboratory at Swarthmore College. As part of an experiment that tested male reactions to familiar and unfamiliar females, a male macaque was placed first with a known female and then with a strange female. Since the familiar female had an infant at that time, to maintain parity the researchers chose an unfamiliar female that also had an infant. In the first case, behavior seemed perfectly normal. But when confronted with the strange female, the male "attacked the infant where it lay clutched to its mother's ventral surface. When the mother struggled and attempted to get away, he pinned her on her back and gnawed at the infant" (Thompson 1967, 18). The attack was notable because it appeared totally uncharacteristic of that male; it was immediate, intensely violent, and unstoppable. Having described the attack, Nicholas Thompson cited Sugiyama's 1967 account of infant biting in the wild and suggested that from "the point of view of sexual selection, such attacks by males taking over a group are very advantageous to the male because, as Sugiyama points out 'loss of an infant has the effect of advancing the estrus of the female'" (1967, 19). This was the first time in the literature that sexual selection and infanticide were directly associated.

Langur Variability (1968–74)

By 1968 the variability in the social behaviors seen both between and within primate species was clearly acknowledged, and it appeared that langurs could well be more variable than most species. In fact, as the final sections of chapter 2 demonstrated, in many ways they were coming to be acknowledged as the "type case" of primate variability. However, it was emphatically not clear at this point whether this variability occurred naturally or happened because some langurs lived in uncharacteristically stressed conditions. Phyllis Jay's position on the matter, published in 1968, makes this explicit. In this edited collection, intended to stand as a current summary of human knowledge of the nonhuman primates,[26] she dismissed the supposition that "the behaviour of a whole species might be meaningfully described on the basis of a small sample from a single location, and even that the behaviour of a genus might be inferred from the behaviour of such a sample" (1968, 173) as an unfortunate consequence of the earlier presumption that primate behavior was largely instinctive. Once primatologists came to realize how much learning was involved in the social behavior of primates, she argued, it became impossible to deny variability. It also became essential to expand the number and range of behavioral studies as far and as fast as possible, even to areas where the environment was being irrevocably altered by humans, in order to record the range of such variability (174). In the case of langurs, however, she noted that Kenji Yoshiba's account of the Dharwar langurs, published as part of the same collection, represented "an extreme in the relationship of a normal social group to extra group males, and it is not clear at present whether this is a common langur pattern, or a local situation, perhaps caused by extreme overcrowding" (178). In other words, the explanation for primate variability can be found in local ecological conditions—but some conditions are more conducive than others to the expression of common patterns of behavior.

Continuing the ecological theme, Yoshiba's article provided an explicit comparison of the environmental conditions at the three sites where langurs have so far been studied, focusing on the conditions of observation, the site ecosystems, the human influence on the natural environment, the population density, and the social structure that appeared characteristic of each area. It was immediately apparent that the population density of Dharwar was comparatively very high—at least thirteen times what Jay observed at Orcha (Yoshiba 1968, 223–24)—and that while multimale troops seemed to exist at both Orcha and Dharwar, by far the majority of troops at Dharwar followed the one-male pattern.[27] The other key differences between social behavior at the sites seemed to follow as logical consequences of this high population density: groups were in much closer daily contact at Dharwar than at the other sites, and violent social change accompanied by the death or disappearance of infants was seen only at Dharwar. Yoshiba emphasized that the full circumstances surrounding infant killing remained unclear: infant killing was not always seen in situations of social change; it was a behavior exhibited by both troop males and males from the male group; and the reasons for males' attacks on infants were not known. In relation to the issue of male motivation, in at least one case, the "infants born *after* the change of the leader male were never attacked by the new leader although it was certain that they had no genetic relationship to him. . . . [I]t was observed, however, that a mother that had lost an infant was soon in estrus and copulated with the new leader, whereas a lactating female seldom was sexually active" (1968, 236). Like Sugiyama's, however, Yoshiba's explanation for infant killing was oriented toward the immediate motivations of the male, not toward the overall reproductive success of an individual animal. His suggestion was that the attack on infants is connected with high sexual activity and "related to a stronger emphasis on some of the behavioural tendencies common to all male langurs" (1968, 242). Yoshiba explicitly concluded that the adaptive value of infant killing was not evident—but he was also concerned to emphasize that "presently available data are not sufficient to warrant detailed intertroop comparisons of langur social behaviour" (1968, 242). More research was needed before definite conclusions could be drawn about langur behavior.

Over the next five years, reports from more than half a dozen new studies of langurs appeared in the literature, confirming and, if anything, emphasizing langur variability in relation to social change, group structure, aggression, and infant killing. Irwin Bernstein reported on the behavior of *Presbytis cristatus* at Kuala Selangor in Malaysia between February 1965 and June 1966, finding that the one-male troop structure seemed characteristic of this species and that groups appeared to defend at least a core area of their territory with vigor. He suggested that the one-male pattern and the extremely unequal sex ratio within the groups could be explained by Sugiyama's proposals, and that the "almost complete lack of juveniles in E troop . . . the small number of juvenile males in other lutong troops, the association of immature males with the adult male and the intertroop relationships described could all be accounted for by the mechanisms Sugiyama describes" (Bernstein 1968, 13).[28] Frank Poirier completed a study of langurs in the Nilgiri Hills of southern India that ran from August 1965 to August 1966, and found that, again, troops tended to follow the one-male pattern. He explored several reasons for this without reaching a clear conclusion. He rejected the idea that fewer males than females were born, was scornful of the supposition that mutual antagonism meant males could not tolerate each other,[29] and pointed out that the only instance where he saw contact between strange males and a bisexual troop was relatively peaceful, suggesting that Sugiyama's "analysis might not be applicable to the Nilgiri langur situation" (Poirier 1969, 27). Troop change did occur, but it was not accompanied by the aggression seen at Dharwar. He concluded that, unlike other species (baboons or chimpanzees) where social groups either are open to new members or are stable and closed, langurs "seem to be polarised at both extremes," with the northern Indian langurs showing stability and the southern flexibility (1969, 37).[30] Hafeezur Rahaman watched langurs for a short time in 1971 in the Gir Forest in Gujarat State, but although it was possible to recognize the existence of multimale troops, visibility there was too poor for observing social behavior (Rahaman 1973, 297). Christian Vogel began comparative observations of langurs at two sites (the

Kumaon Hills near Bhimtal and a wildlife sanctuary in Rajasthan) between September and November 1968, specifically to examine how strongly behavioral variability was related to the local ecology (Vogel 1971), and he argued that the one-male/multimale troop structure of langurs should "not be seen as static forms of organisation; they are rather, to be interpreted generally as the various stages in the dynamic development of the group. . . . In a few years, a 1-male group can evolve into a group with two or more adult males and the structure can change again, in that a male drives his rivals out of the group forming once again a 1-male group" (1973, 357). Langur society appeared to be characterized by both chaos and continuity, and there was little consensus at this stage on what mechanisms might produce this apparent situation.

Infant disappearances—and crucially, infant killing—were also seen at two other sites. In 1971, S. M. Mohnot reported the strange case of group B26 at Jodhpur, in the state of Rajasthan, where langurs had been under observation since July 1967 and where no multimale troops had been observed. Toward the end of June 1969, some unknown force had devastated B26, which previously had contained over eighty animals. Only eleven animals known to belong to the group could be found alive; and these remaining seven adult females and four infants were then subjected to waves of attacks, which Mohnot charted by date and time, from the male group (M27) that also inhabited that area of the site. After the second attack, one male remained with the remnants of B26 and "in the next few weeks he killed all the four infants" (Mohnot 1971, 185). As Sugiyama did, Mohnot gave a minute-by-minute account of each encounter between the troop and the male group, but in this case the observer could record direct attacks on infants that led to infant death. Of the four, the male was seen to bite three, and they died shortly thereafter. The last disappeared at about the same time, and an infant born toward the end of January (five months after the new leader had established control over the group) also went missing. Mohnot believed both had been killed by the new leader, but his explanation was rather different from Sugiyama's. Mohnot did not accept that the infants were killed so the male might have sex with the females, since the male was not seen to copulate with the infant-deprived mothers; instead, the females mated with other males in the M27 group. Mohnot suggested that "simultaneous sexual excitement and enragement [at the ongoing attacks from the males of M27] made the new leader very aggressive, and this might have been the stimulus behind infant killing. His post killing behaviour supports this presumption. Each time after the killing, the leader looked relaxed and became mild" (1971, 196–97). Infant killing, then, was a way for the adult male to dissipate aggressive tension in a relatively "safe" way—for the male, if not for the infants.

In the second case, infant killing was not seen but was strongly suspected. Rasanayagam Rudran studied purple-faced langurs at two sites in Ceylon (Sri Lanka), and found a pattern of one-male troops and male replacements that appeared very similar to what Sugiyama had recorded for Hanuman langurs. However, the study, which lasted from February 1968 until June 1970, was not continuous (the sites were visited alternately each month), so that "observations on social changes could only be periodically checked through several days' observations every other month" (Rudran 1973, 170). Nevertheless, it was possible to note patterns in social change. At both the study sites (Polonnaruwa and Horton Plains), despite clear differences in population density, the one-male troop structure accompanied by a number of all-male groups was overwhelmingly dominant, with no apparent changes in the troop membership other than those produced by infant births and deaths and the replacement of the adult male. At least five cases of male replacement were thought to have occurred at Polonnaruwa, and with one exception "all other infants which were present in troops which underwent replacement were found to be missing from their parental troop as well as the study area after the social change . . . they were thought to have died during the process of adult male replacement" (1973, 172). Male replacements were not seen at Horton Plains, despite intense aggression between adult males and male groups. Like Sugiyama, Rudran focused on explaining the maintenance of the one-male troop structure, arguing that "male replacements appear to be the most common mechanism by which this structure is maintained" (1973, 177) and at Polonnaruwa were probably responsible for the complete lack of subadults and juveniles in troops, since these individuals would be expelled along with the deposed male. Nearly a decade after Sugiyama's initial observations, Rudran used the greatly expanded langur literature to develop a scheme of langur social structure postulating that "male replacements are more widely prevalent among langur populations than has been previously assumed and . . . are considered more effective in maintaining the one-male troop structure and regulating population growth than other mechanisms" (1973, 180). The frequency of male replacements was directly related to population density. Where levels were low, male tolerance of familiar infants allowed a multi-male troop structure to emerge. Where density levels were high, constant contact between one-male troops and male groups led to frequent male replacement, thus preventing the development of a multimale troop and causing the expulsion or destruction of entire age-sex classes within that troop. Here again the focus was on the structure of the troop and the population, not the effects on the individual. Infanticide was merely a means of regulating population density.

Finally, two textbooks published in 1972 summarized the known variability of langur behavior at different sites. Alison Jolly used langurs as the type case of variability in nonhuman primate social behavior and as a warning against hasty generalizations from human to nonhuman primate behavior—after all, if we "cannot extrapolate from a North Indian langur to a South Indian langur," then how much more difficult would it be to link primate and protohominid society (1972, 4)? Her account of langur life, however, tended to treat observed behavior as part of a continuum, with the one-male pattern used as a basis for langur society that is capable, under the appropriate environmental and social conditions, of developing into the multimale structure Jay observed. In addition, Jolly explicitly treated the pattern of troop takeover as part of a sexually selected strategy, arguing that it is in Sugiyama's langurs that one can see "the clearest selection for male dominance. The new harem leader gained exclusive rights to his females. Then he commonly bit young infants after the takeover. Their mothers hung them on a branch to die and came into estrus, ensuring an even higher proportion of the new lord's children" (1972, 181).

After Thompson's 1967 linkage of Sugiyama's observations and sexual selection, this is the second connection between infant biting and sexual selection to appear in print. As previously mentioned, Phyllis Dolhinow included in her edited collection *Primate Patterns*[31] an updated version of her 1965 paper that mentioned the apparent contrast between the behavior of the northern and southern Indian langurs and stressed that the study of behavioral variability was far from complete. However, she was still concerned to emphasize that the "context for langur social life is a stable well-organised troop composed of monkeys of all ages and both sexes. Such a group is the basic unit of the species" (Dolhinow 1972, 204). During her short comparison of the Dharwar observations with her own work, the differences in social structure and behavior are noted and explained by the observation that "the population density at Dharwar is very high" and that it is characterized by a one-male rather than a multimale troop structure (1972, 235). Her conclusion was prescient: langur behavior is undoubtedly variable, and "generalisations written in 1963, revised now in 1969, may be just as much in need of revision in another six years time—providing that there is continuing research" (1972, 238).

Conclusion

In 1974, five years after Dolhinow wrote these words, when Hrdy published her first paper on the langurs of Abu, there was no clear consensus on the characteristic structure of langur society and on individual behavior. Langurs had been found in both one-male and multimale troops, with male groups of varying sizes surrounding them. Sudden, drastic, and aggressive social change had been reported for many of the sites where langurs were studied, and in at least three cases social change had been accompanied by the death, disappearance, or exclusion of infants and other juveniles. A variety of explanations for these events had been put forward. Sugiyama thought they allowed the new male to form stronger social and sexual ties with the troop females. Yoshiba agreed that infant attacks were associated with periods of intense sexual activity but concluded that the motivation behind such attacks was unclear—although he acknowledged that females that still had infants were unlikely to be receptive to sexual approaches and that a mother that had lost her infant would reach estrus much sooner. Both Sugiyama and Yoshiba thought attacks might be provoked by unfamiliarity, since it seemed that males did not attack infants born in the troop during their own tenure. Mohnot argued that it was a by-product of an environment of intense aggression, and a way for the male to dissipate tension. Rudran, like Sugiyama, related it to the maintenance of the one-male troop pattern, while Itani (1972) suggested it was a means of avoiding incest,[32] a possibility echoed by Poirier, who accepted that infanticide might be a means of avoiding inbreeding, as well as easing relations between new males and troop females by allowing them to mate sooner than they otherwise could (Poirier 1974). Dolhinow wondered how the "frequent change of male leadership and the killing of young permits the survival of the langur population" (1972, 366) but suggested that both were the results of the very high population density seen in some areas.

None of these explanations made an explicit connection between infanticide (infant biting, infant attacks)[33] and sexual selection. With one exception—incest avoidance, which was a classic example of an argument drawn from a group-selectionist perspective[34]—they focused on the immediate context of the event—the need to establish a relationship between new males and troop females, the particularly aggressive circumstances of a male replacement, and the ecological conditions that encouraged such replacements. Additionally, for the most part they also tended to focus on the consequences of social change for the troop as a whole rather than its individual members. Dolhinow's attention to troop structure and the nature of the roles that constitute and sustain it has been described in some detail, but this orientation was reflected in the other accounts: remember that Sugiyama's overriding concern was to establish how the one-male troop pattern was maintained at Dharwar. This was not surprising and is related to another characteristic element in the reporting of primate behavior over this period— combining fragmentary observations into coherent narratives structured after the life cycle (literally in relation to individuals, metaphorically in relation to troops). In contrast to primate observations made in later decades, these researchers simply did not have enough information to draw complete pictures of primate life—or conclusions about them. Most of these reports were based on projects that lasted far less than two years, a drastically short time by the standards of early twenty-first-century primatology, and one of the consequences of the limited time available was that (again by the standards of the early twenty-first century), habituation was incomplete. Individual animals were therefore treated as exemplars of age-sex classes within an overall troop structure rather than as individuals whose life histories and interests were known to the observers.

Such matters were shortly to become the central focus of the developing sociobiological perspective in Western primatology. In the meantime, however, it was plain that by 1974 infanticide was accepted as an odd event that occurred from time to time in circumstances that were not well understood and whose adaptive value was not clear. It was not the subject of the intense passion it was to generate after the publication of Hrdy's observations at the hillside town of Abu, which are described and analyzed in the next chapter.

FOUR

From Controversy to Consensus? 1974–84

Introduction

By the time Sarah Hrdy decided to travel to India to study langur monkeys, the key concepts that were to underpin the new sociobiological synthesis were already in circulation, both within the scientific literature and, more specifically, at Harvard University, where Hrdy was a graduate student.[1] However, both at the time and in later interviews, she was insistent that she had gone to India to "test the hypothesis that crowding was responsible for infant-killing among langurs" (Hrdy 1977a, 9–10). Crowding was, of course, the hypothesis favored by influential American langur watchers such as Dolhinow, and such an approach made sense: not only was infant killing rare, it was associated with only three sites in India and Ceylon,[2] and at least two seemed to be characterized by unusual socioecological circumstances. In addition, there was laboratory evidence that severe social crowding would produce pathological behaviors in rats—behaviors that included infant killing (Calhoun 1962a, 1962b).[3] Given that the most famous infant-killing site—Dharwar—also exhibited very high langur population density, it was logical to assume these two curious characteristics might be related.

However, when Hrdy began to publish the results of her studies at the hillside town of Abu in southwestern Rajasthan, it was clear that she had abandoned the crowding hypothesis, along with the assumption that infanticidal sites were in some way unusual. Quite the contrary: she argued that infanticide, at least for Hanuman langurs, occurred "under conditions that now must be considered normal for this species" and was an example of "a reproductive strategy whereby the usurping male increases his reproductive success at the expense" of other animals (1974, 20). In the articles on this topic that she published over the next few years, Hrdy went further, placing the act of infant killing at the heart of her attempt to develop an unified and universal interpretation of langur social structure and even arguing that the ever-present danger of primate infanticide might have been responsible for one of the most unusual aspects of humanity itself—the continuous sexual receptivity of the human female.[4] This bold attempt by a young female graduate student to reorder the understanding of langur and primate social structure was bound to encounter hostility. It is also clear that the fierce debates Hrdy's arguments provoked were exacerbated because they were firmly grounded in sociobiological theory, itself fast becoming the target of hostility, suspicion, and outrage in the years following the publication of E. O. Wilson's *Sociobiology* (1975a).[5] However, although the controversy surrounding langur infanticide was undoubtedly part of the much broader dispute over the development of sociobiology, sociobiology was not the only source of the controversy. The infanticide debate, despite its wider significance, initially was closely framed by the specific contexts within which science in the field was constructed and constrained, contexts initially explored in chapter 1.

This chapter will examine the way Hrdy's thesis about the nature of langur social structure and the role of infanticide in mammalian social systems was first presented to and received by the wider primatological and intellectual communities. It will consider the ways Hrdy chose—and occasionally was forced—to conduct her investigations at Abu and will show how these decisions influenced not only her results but the way she presented those results to her audiences. I will show how audience responses—both critical and laudatory—were in turn informed by descriptions of the contexts within which infant killing had been reported, and I will examine how the range of critical reactions to the sexual selection hypothesis narrowed over the next decade. These questions must also be considered in relation to the issues that had bedeviled primatologists from the inception of the modern period of primatological research and that I identified in the introduction as the conditions that produce the "fieldworker's regress." In other words, responses were affected by the need to define what was to count as "natural" primate behavior in a fast-globalizing world where humans were rapidly altering local environments; by the difficulty of managing the close relation between human and nonhuman primates and avoiding overt or implicit anthropomorphism; and finally, by the problems inherent in operationalizing the relation between theory and methodology and the interconnections between observation and experiment, evidence and authority in a field context. All these issues had been raised frequently throughout the history of primatology in the twentieth century, but what was unusual in the context of infanticide was the way participants in the controversy treated them as novel, relevant only to infanticide. Black boxes were being flung open with a vengeance.[6]

Abu and Adaptation

From June 1971 until June 1975, Hrdy carried out her graduate research at Abu. She made five visits to the hillside town, each lasting two to three months, and the work she did there was repeatedly scrutinized as the infanticide controversy developed. Six troops lived in the vicinity of Mount Abu,[7] along with a number of male bands, and she was able to obtain over 1,500 hours of observation on these animals (Hrdy 1977a, 57). However, since her interest lay in recording the events surrounding infant killing and (eventually) the reproductive strategies of both sexes, she made no attempt to divide her attention equally between the troops or to systematically observe each individual in turn. Instead, she chose to focus on troops where events were occurring that were likely to have an immediate impact on reproductive matters—troops where an estrous female or extratroop males were present, for example. For this reason the Hillside and Bazaar troops dominated Hrdy's accounts of the langurs at Abu, since of the nine male troop leadership changes that occurred there, with their attendant social upheaval and infant mortality, seven took place in these two troops alone.[8]

Hrdy described the key events over this five-year period in three publications (1974, 1977a, 1977b).[9] In July 1971 she made her first contact with the Hillside troop, noting that it contained one adult male, six adult females, two juveniles, and six infants. When the troop was next seen, in August 1971, not only was a different male in residence, but all six infants and one female were missing. Local inhabitants told her that "an adult male langur had killed two infants" within the territory known to be used by Hillside (Hrdy 1974, 27). In the article that appeared in the peer-reviewed literature, Hrdy recorded dispassionately that this information, together with its similarity to the reports from Dharwar and Jodhpur, "led [her] to assume that the new male had killed at least two and very probably all six infants" (1974, 28). In *The Langurs of Abu* she went into rather more detail about her shock and surprise at this turn of events:

> Despite Shifty's replacement of Mug, despite the fact that all six infants were missing, despite reports by two local people who had seen an adult male langur kill infants in the Hillside troop's home range, I grasped at straws. I spent a whole day trying to convince myself that this was a different troop, one

> without infants, which had somehow materialised out of the torrential rains and thick mist of that monsoon month. But the longer I peered through the mist at those rain-soaked, skittish females, the more I realised they were, unmistakably, Bilgay, Itch, Harrieta, Oedipa, Pawless and Sol. The seventh female and the six infants were never seen again. (1977a, 243)

Hrdy left Abu the next month (September 1971) and was unable to return until June 1972. When she did, it was to the news that Shifty, the male that had taken over the Hillside troop the previous August, had in her absence transferred to the Bazaar troop and had killed two infants there. This information was passed to her by a local amateur ornithologist whose roof the Bazaar troop used as one of its nighttime resting places. After observing the troop and comparing its current composition with that recorded in the previous year, Hrdy was able to confirm that not two but three infants were missing, along with two juvenile males, an adult female, and the three males that had previously belonged to the group. Based on these demographic changes, on the information from her local informant, and on the geographical location of events, she concluded that "Shifty was responsible for the deaths of the missing infants" in the Bazaar troop (1974, 30).

The battle between Shifty and Mug for control of Hillside had not ended, however. Throughout her observation in the summer of 1972, the two males competed for residency in the Hillside troop—or perhaps "competed" is the wrong word, since it appeared that Mug would swiftly retire from the vicinity whenever Shifty approached. However, when Mug was with the troop he "stalked B-6 (Hillside) mothers with infants" (1974, 35), infants that had presumably been fathered by Shifty. Nine attacks were made on these infants during the 1972 study period, none ending in the death of an infant, although one (Scratch) was severely wounded. Scratch lived through this attack, but by the time Hrdy returned to Abu in spring 1973 he had disappeared, along with two other Hillside infants, and an adult male had been seen to kill an infant two days after Hrdy's departure. Based on the known composition of the troop and on Mug's behavior, she assumed he was responsible for two of the deaths: the third was more questionable, especially since five other males were also now periodically present in the troop. Although Mug seemed dominant over them, there was no way of knowing if he was able to monopolize access to the females, and in any case Shifty was still able to disperse all six males during his periodic visits to his erstwhile troop. In his absence, though, the males continued to attack the only young infant left (Pawla). Just to add to the confusion, Hrdy later realized, when watching the films she had made of these attacks, that Mug might actually be *defending* the threatened Pawla against the other males.

By the time of Hrdy's next visit to Abu in December 1973, however, Pawla had disappeared, along with the extraneous males, and Mug appeared to be the sole resident male of Hillside. But at some point between January 1974 (when Hrdy left) and March 1975 (the beginning of the final study period), Shifty disappeared completely. Mug was now the resident male of the larger Bazaar troop, and one member of the group of five males from the spring of 1973—Righty Ear—had taken his place in Hillside. Then, only a few weeks after her arrival, local informants told her that an adult male had killed an infant in the territory of the Bazaar troop. On investigating, she found that Mug had vanished and that in his place was Righty Ear. One female was carrying an infant's corpse, and Righty Ear went on to subject the six remaining mothers and infants to repeated assaults over the next two months. Although one infant disappeared, the other five were alive and well when Hrdy finished her observations in June. As to Hillside, when another researcher (James Malcolm) visited the troop in October 1975 Righty's place had been taken by the male Slash-Neck, and a table included in the text (Hrdy 1977a, 267) indicated that two infants were missing, although the circumstances of their disappearance were not recorded.

In total, Hrdy had recorded nine male replacements over her study at Abu, accompanied by the death or disappearance of eighteen infants. In combination with the observations at Dharwar and Jodhpur, she argued that thirty-nine infants were now known to have disappeared or died during male leadership changes, and together with a number of observations of similar events in other primate species, she treated these numbers as the database against which to test her sexual selection hypothesis. Her critics, however, were soon to take direct issue with the resilience of these data, particularly in relation to *how much* of an apparently infanticidal attack had been observed and recorded by the reporting scientist, as well as how far infanticidal sites could be treated as places where one might see *representative* examples of langur behavior.

The State of the Site

Abu was, at this point, a town with about eight thousand settled inhabitants as well as a transient population of pilgrims and tourists who lived in close company with the langur troops Hrdy studied. This association was critical not only for the success of Hrdy's research at the site, but also for understanding the way other primate researchers responded to her accounts of what she had seen there. Historically, primatologists preferred to work at sites that had remained relatively undisturbed by human interference, so as to present their work as reflecting the "natural" behavior of nonhuman primates; in contrast, work conducted where the local human population had influenced of animals' activities, either through environmental changes (agriculture or deforestation) or by direct interaction (provisioning), was potentially compromised by that influence. That many of the langurs of Abu either lived within the town, scavenging for food from the human population, or were directly fed by travelers, tourists, and pilgrims might be seen to invalidate any argument that the behaviors Hrdy recorded reflected centuries of langur evolution.

Additionally, despite the value primatologists placed on long-term continuous observations of primate populations, Hrdy's work was based on short-term, interrupted periods of data collection. This was unavoidable: the tense political relations between the United States and India during the early 1970s made it difficult for American researchers to get visas to remain in the country longer than three months.[10] As she put it:

> We were Americans, the U.S. had tilted toward Pakistan. Local people and scientific colleagues were very helpful, but at a bureaucratic level we were treated in Rajasthan as if we might be spies. . . . I was there on a tourist visa, funded as a tourist by my mother in the first however many years of the study until the Smithsonian and NSF got involved, at which point, because the research would be formally funded, it was essential to have official research permission. It wouldn't do to have to leave the country every three months, or else be arrested, and even following all the rules, to not necessarily be guaranteed that I could return or stay longer. So, to say that this influenced the research is to put it mildly. There were no other Americans working on langurs in India at the time, and this is the reason.[11]

This placed severe constraints on her ability to watch the langurs and was the reason for the periodic breaks in her research record. Moreover, she was observing not one but seven bisexual troops, meaning that continuity of observation on any one troop could be abruptly disrupted by the need to monitor what was happening elsewhere—again, a problem when the subject of study is not the regular daily activities of troops but the unusual events surrounding reproduction and infant killing. For this reason, in each of the three publications that heralded the onset of the controversy surrounding infant killing in langurs, Hrdy not only made a virtue out of necessity but also killed several birds with one stone.

Rather than treating the close association with humans as a problem for the research, she portrayed it not only as a solution to the unavoidable hiatuses in her observation records, but also as a positive benefit. The presence of the local human community meant that much of the preparatory work necessary for the study of primate social behavior had already been done for her. Not only were there detailed maps of the area, but the problem of habituating the animals to her presence had been solved; the langurs were used to receiving food from humans, and even those troops that were not often fed by the local population swiftly became tolerant of her passive presence, even though the religious beliefs that protected langurs from harm were slowly eroding.[12] Their dependence on human food meant she could make close observations of social behavior soon after the study began.[13] Regular feeding maintained habituation of the langurs even when she was absent and also enabled local informants to make constructive contributions to the records of behavior at the site. Notably, in her first description of the conditions of her research at Abu, Hrdy stated that "unusual occurrences (such as leadership change or infant killing) were sometimes witnessed by local people. Where reports from reliable informants coincided with what I knew about troop locations and changes in troop compositions, they were accepted by me as facts" (1974, 24).[14] In this way Hrdy managed to present as serendipity one of the biggest pitfalls to any acceptance of her work as a representative account of the evolved behavior of langurs, given the constraints imposed on her fieldwork.

But her account's dependence on local reporting rather than personal observation of infant death and the intertwining of langur and human life at Abu were to be critical to the development of the debate, as we will see by examining the responses to Hrdy's publications. As she acknowledged,

> I knew the evidence [at that stage] was shaky, incomplete, skimpy, which is why I called it a hypothesis that generated predictions, and you know, nobody then was saying, this is a fait accompli, this is it. But the other thing, as soon as the other evidence started to come in, and the patterns were the same, and the predictions were being confirmed in the way they were, we could actually predict who was going to be attacked by whom, under what circumstances [then] I felt like I had a pat hand.[15]

In the absence of data that could be shown to provide overwhelming support for Hrdy's adaptive explanation for langur infanticide, she constructed a powerful narrative model of langur social life. This model not only included and accounted for the observations made at the sites where infant killing had been reported, but also incorporated the results of research at sites where langurs were reported to interact peacefully. In addition, it proposed further research strategies for testing this hypothesis and exploring its ramifications for understanding primate social behavior. In the classic sense, this put forward a whole new primate paradigm: a paradigm that focused on identifying consistent and predictable patterns rather than on documenting primate variation.

Langur Life Cycles and Infanticidal Type Cases

As the previous two chapters demonstrated, a decade and half of work on primate social organization had described a bewildering variety of structures both between and within primate species. In particular, there appeared to be a stark discrepancy between northern and southern Indian langurs in the number of males that could typically be said to reside within a bisexual troop. At Orcha, Kaukori, Polonnaruwa, and Raipur, troops tended to contain more than one male, whereas at Dharwar, Jodhpur, and Abu—the places where infanticidal events had been reported—most troops had only a single adult male. This clear division had been a key factor behind the assumption that infanticidal sites were atypical, characterized by an unusual socioecology that could also account for infant killing. Hrdy's innovation was to develop a cyclical pattern of langur troop structure development, similar to what Sugiyama had proposed in 1964 but using sociobiological theory as the driving mechanism. This model enabled her to make sense of her own observations while integrating them with the apparently contradictory results reported by some of her predecessors reported.

She did so by treating the accounts of langur social life as isolated instances of a wider pattern that could be discerned by tying these fragments together into a "hypothetical scenario of the life cycle of a langur troop" (1974, 25). At any given point, she argued, a langur troop will consist of a core group of closely related females and at least one adult male. Those males not found in bisexual troops live in all-male bands and range over the territories of a number of troops. When bands and troops encounter each other the males become aggressive, and in certain circumstances the male band will oust the troop's adult male. One of the remaining males will eventually succeed in driving off his erstwhile allies, along with any male juveniles, and will go on to kill any dependent infants. This, for Hrdy, represented the *initial stage* of langur troop development, where the clock is reset: the old male is evicted, his immature offspring are eliminated, and the new male is left alone in the troop with only the adult or near-adult females remaining. Destroying their unweaned infants will trigger the resumption of estral cycling in the adult females, who will mate with the new male and become pregnant by him, causing the troop to enter a *growth stage* where the population will increase rapidly, a process that may continue for several years until the troop enters the *mature stage,* where all age grades are once again represented. Hence the observations from sites where this rapid male replacement did not occur were accounted for: the observations at Kaukori and Orcha were presumably of troops in this *mature stage,* where males "growing up in a troop may ease into possession of it" (1974, 27). However, in areas where there were many males living outside the troop—for example, where population density is high, as at Dharwar—encounters between male bands and bisexual troops leading to takeover happened more frequently, and social change was therefore more likely to be observed.

This account of langur troop structure was fundamental to Hrdy's interpretation of infanticide as an adaptive strategy, and it rested on two key pillars. The first was the nature of langur reproductive physiology. A female langur with an infant will not return to fertility until that infant is weaned: while the female is still nursing, the male cannot impregnate her with his own offspring. This was key to Hrdy's account: while Sugiyama had suggested that infanticide might occur as a consequence of male competition for females, he did not take the further step of placing that competition in its reproductive context. He argued that males might kill infants to avoid a delay in a female's return to sexual *receptivity,* while Hrdy focused on the female's return to *fertility.* The significance of this can be found in the second pillar Hrdy's account rested on: the implication that langur troops are closed social and reproductive units. If only males resident in a troop have the opportunity to mate with troop females, then gaining access to a troop is essential to male reproduction.[16] Where troop takeovers are frequent and where a male's residency in a troop, and hence his ability to reproduce, may be short, killing unweaned infants enables a male to maximize his reproductive success. It is relatively sure—if only troop males have reliable reproductive access to females—that he is not only eliminating the progeny of a rival but accelerating the production of his own offspring and thereby maximizing the chance that they will have been weaned, and thus be able to survive, when he himself is replaced by another rival. In circumstances of intense male-male competition for females, as at the crowded sites of Dharwar and Abu, Hrdy's narrative provided a compelling logic for an apparently aberrant behavior.

But despite the rhetorical power of the model Hrdy has produced, it remains, as she acknowledged, *hypothetical.* The outline of troop development was one she had constructed from individual observations, not a process she had observed in its entirety. Its plausibility comes from her ability to show that it rests on recorded examples of natural facts: each stage is illustrated by actual events at Abu and elsewhere, organized not chronologically but according to the demands of her narrative, cyclical structure. At the heart of this structure lies the act of infanticide: just as her adaptive account of infanticide depended on accepting this model of the langur troop's life cycle, so the acceptability of the model depended on acknowledging infanticide as an evolved mechanism. She used a corresponding narrative strategy to explain the fragmented record at Abu: one makes sense of incomplete observations by placing them within a logical pattern that is, again, never seen in its entirety. The "type case" of langur infanticide was the one Sugiyama reported for the thirtieth troop at Dharwar, involving a male invasion, the eviction of the previous leader and (eventually) all but one of the male intruders, followed by the killing of all the troop's infants. Although events at the other infanticidal sites did indeed show striking similarities to this pattern, they did not follow it slavishly.[17] As we have seen, Hrdy herself never actually saw an adult male kill an infant, and her account of this repetitive cycle of competitive usurpation was based on a series of reconstructions of events in the light of changes in troop composition and the reports of local people.

Given the paucity of the observed record, how could Hrdy justify making the extremely significant claim that infanticide occurs repeatedly and "under conditions which now must be considered normal for this species" (1974, 19–20)? She managed by using her observations merely as the basis for her narrative account of langur social structure; events at Abu are treated as illustrative of the pattern she claimed could be shown to be representative of all langurs. They were not pathological because the attacks were "highly goal-directed and organised"; because they seemed to be carried out according to "roughly accurate assessments of paternity"; because females seemed to have evolved strategies for avoiding infanticidal attacks; and because "high levels of aggression [among langurs] have been ongoing for at least 130 years" (1974, 45).[18]

The Narrative Construction of Langur Life

This last point regarding the historical precedent for observations of infanticide became a critical resource when Hrdy took her hypothesis to the wider audience of *American Scientist* (1977b). The article deserves a detailed examination, since it represents the public eruption of the infanticide controversy and was the starting point for a series of interchanges between controversy participants in this journal. While her previous accounts of infanticide had concentrated on langurs, this article went much further, arguing—as the title indicates—that infanticide could now be seen as a much more widespread "primate reproductive strategy" (1977b, 40). In the first section Hrdy outlined her model of langur social life, where the females represent the stable core of the troop while the males leave or are driven out on sexual maturity, returning only if they are "successful in invading a bisexual troop and usurping resident males" (1977b, 40). This description, given in a journal oriented to an interested, knowledgeable, but nonspecialist audience of scientists and laypeople, clearly gave the impression that this view of the sexually demarcated constituent elements of langur society—including the importance of direct conflict between males—was held by most modern workers.

However, almost at once Hrdy signaled the possibility of controversy by discussing the historical precedent for her work. She acknowledged that the generation of researchers who preceded her in the field did not consider aggression a crucial component of langur life, but she proposed that there was a clear, though unfortunate, explanation. The "*social* scientists" who began the "modern era of primate studies" had been "profoundly influenced by current social theory"—specifically, the integrative theories of Alfred R. Radcliffe-Brown (1977b, 41; emphasis added). Precisely because of this preexisting commitment to a view of social life that emphasized cooperation as central to community life, these early modern researchers had taken it for granted that "primates behave as they do for the good of the group" (41). As a result, they had dismissed as inaccurate anecdotes the accounts of langur aggression dating from the nineteenth and early twentieth centuries. Hrdy's implication here was obvious; if these later researchers' vision had not been clouded by their theoretical orientation, then, like their predecessors and their successors, they too would have realized the importance of aggression in a stable langur society.

Again, Hrdy's account of her empirical observations acknowledged breaks in the continuity of her research record, but now her account of her observations at Abu was far more tightly coupled with a powerful narrative of the reconstructed events there. As in her earlier publication, she acknowledged that her periods in the field were short and separated by up to a year, and again she drew openly on others' observations to fill these gaps,[19] reporting not only her records of animal behavior but her assumptions about what happened when the animals were out of sight. In particular, she did this with the infant attacks, describing how the victims "disappeared and [were] presumed dead" (1977b, 44). A dependent infant separated from its mother has no chance of survival—but Hrdy was plainly ascribing the cause of death to a male attack rather than maternal abandonment. Her inference and her implication were that social change in langur troops is frequently accompanied by infant death—and she managed to render the potential empirical weaknesses of her account almost irrelevant by her theoretical justification. It no longer matters that she did not see males killing infants, because she has clearly shown the audience how males *ought,* in certain circumstances, to kill infants. In a context where males have only brief access to females (as would happen in Hrdy's cyclical model of langur troop growth and development), "infanticide permits an incoming male to use his short reign more efficiently than if he allowed unweaned infants present in the troop at his entrance to survive, to continue to suckle and thus to delay the mother's next conception. . . . Once infant killing began, a usurper would be penalised for *not* committing infanticide" (1977b, 45; emphasis in original).

A noninfanticidal male would risk that his own offspring would still be unweaned and vulnerable when he was ousted from the troop. Not all takeovers are accompanied by infant death,[20] nor *must* they be: the essential point Hrdy must make is that attacks on infants occur *only* when stranger males—which could not have fathered the infants—enter the troop. This timing and the targeting of a specific category of infants were the key reasons she rejected the idea that this behavior was pathological—instead, it was "highly goal directed" (1977b, 46).

The organization of this article and the nature of the claims made in it tended to render Hrdy's competition hypothesis unproblematic in the following ways. First, her account of the current state of research, written for a nonspecialist audience, presented a particular and potentially idiosyncratic model of langur society as commanding general acceptance. Second, she effectively sidelined her potential critics by implying that their work was outdated[21] and their perceptions were colored by their preconceptions about the basic nature of society. Third, she eradicated potential alternatives: she portrayed langur society as made up of closed social units where there can be many females but only one male. Therefore, in order to breed, males must force their way into mixed troops. Her development of this powerfully cyclical narrative of takeover and replacement was given far more space than either the apparently stable social structure of the other troops she observed in the district or the discussion of the problems her model faced because of her limited field observation time. Fourth, she placed her account of langur infanticides firmly in the context of the wider primate literature and the comparative data: rather than isolated instances of odd behavior at certain field sites, they became examples of a pattern seen in "every major group of primates" (1977b, 46), including humanity. Finally, she closed her account with an array of further research questions that must now be answered, thus rendering it moot whether her interpretation of the (reconstructed) events at Abu was accurate: primatology must now move on to examine the details of *how* the individual primate is able to maximize his or her reproductive success in a context where infanticidal strategies are an option.[22]

Reflections and Refutations: Reordering the Narratives

Indeed, by 1977 the topic of infanticide had begun to arise fairly regularly in the primatological literature, and not just in relation to langurs. For some authors, Hrdy's adaptive account of infanticide as the product of sexual selection was profoundly satisfying, closely fitting their own experiences. Roonwal and Mohnot (1977), in a review of the literature on the monkeys of Southeast Asia, treated infanticide both as characteristic of Hanuman langurs (rather than of specific field sites) and as a product of evolutionary selection: they accepted Hrdy's argument that the existence of female counterstrategies and the highly goal-directed nature of male attacks on infants made nonsense of the social pathology hypothesis. Similarly, Struhsaker (1977) reported harem male replacement accompanied by infanticide in redtail monkeys—the first to be seen in African rather than Asian monkeys—and pointed out that infanticide occurs *only* where the death of the infant would hasten the mother's return to estrus and where males can be tolerably certain the targeted infants are not their own. Again, this meant that infanticide did not always occur even where observers might expect it: the cost of accidentally killing one's own offspring was so high that it paid to err on the side of caution. Outside the Primate order, Bertram (1976) discussed infanticide as an unequivocal example of a strategy male lions used to improve their own reproductive output at the expense of another. In each of these examples, the crucial factor behind the writer's acceptance of the sexual selection hypothesis appeared to be that infanticide occurs *only* when certain conditions are met: where the male is not familiar with the female, or where he has his first opportunity to mate with the female and the infants involved are extremely unlikely to be his own.

Other researchers, while willing to accept that infanticide might be adaptive, were less keen to adopt the sexual selection hypothesis. Sugiyama, for example, accepted that that the pattern of male attacks on a bisexual troop, resulting in male replacement and the killing of infants, was now accepted as a "common phenomena" for langur populations. However, he continued to treat it as a mechanism for maintaining a one-male troop structure—as the process through which multimale troops are returned to single-male status (Sugiyama 1976, 274). Wolf and Fleagle also adopted this perspective as a result of their research on langurs in Malaysia (1977). Angst and Thommen (1977), in a review of reports of infant killings throughout the Primate order, took a slightly different tack. Like the supporters of the sexual selection hypothesis, they argued that infant attacks are liable to occur when certain conditions are met—specifically, where a male either is new to the group itself or has only recently achieved dominant status. However, their interpretation of this behavior, like Sugiyama's, focused on the immediate social context of the attacks: they argued that infant biting is a special kind of aggression that is provoked by the incomplete integration of the new male into the group. The death of the infant hastens its mother's return to estrus, thereby accelerating the bonding between the new male and the troop's females. In this way, they suggested, although the behavior is initially disruptive, it nonetheless hastens the reformation of the troop around the new male. Sexual selection may encourage male killing of infants, but they believed there was too little information to draw any clear conclusions. Similarly, Goodall's review of the five chimp-caused infant deaths at Gombe did not unequivocally support the sexual selection hypothesis, although she treated the behavior as at least potentially adaptive.[23] Many primatologists, while wary of accepting the sexual selection hypothesis in its entirety, were cautiously willing to acknowledge that this profoundly disruptive act might in fact be the product of selective pressure: that is, an evolved strategy. Others, however, found such a suggestion abhorrent.

Normal Monkeys?

Phyllis Dolhinow, the first researcher to study wild langurs and hence by implication one of the misguided "social scientists" of Hrdy's *American Scientist* article, reacted with horror to these interpretations of the events at Abu and the literature on primate infant killing in general. Her objections were manifold but focused mostly on how far Hrdy's animals—and those at other sites where infant attacks had been reported—could be considered "normal monkeys."[24] These sites, she stressed in a letter to the editors of *American Scientist*, were overcrowded and suffered from high levels of human interference, which had drastically affected the animals' behavior.[25] The best proof of this, for Dolhinow, was the high infant death rate itself: rather than being the direct result of evolutionary pressures, this was clear evidence that something was profoundly wrong. The "normal" pattern of male replacement, she argued, could be seen at a study site in Nepal where—in the absence of disruptive human influence—changes in the male population of troops occurred regularly and peaceably.[26] In response, Hrdy took direct issue with this issue of "normality," questioning both the usefulness of the concept itself and its application to specific field sites.[27] Humans and langurs, she insisted, have lived in close association in India for many centuries: How then could one argue that human presence inevitably disrupted patterns of langur behavior? Moreover, not only had Struhsaker's work in Uganda shown that infanticide occurred even at sites without human presence, but even if one accepted that Abu was a "disturbed" site, infanticide still occurred *only* in troops where male takeovers occurred.

It is clear that both women, despite Hrdy's evident unease with the term "normal," are attempting to position their own sites (or those of their graduate students) as the most representative of the spectrum of langur social behavior, a point picked up by Glenn Hausfater, who pointed out that "the fact that infanticide did not occur at the study sites mentioned by Dr. Jay does not constitute evidence that infanticide also did not occur at the various study sites reviewed by Dr. Hrdy. Likewise, the fact that isolated cases of infanticide have been reported for numerous primate species does not in any way indicate that . . . infanticide as a behavioural trait is somehow latent in all primate species" (1977, 404).

In other words, both Dolhinow and Hrdy were in danger of falling into the double-baited trap that had consistently occurred in primatological debates and would continue to do so: the inadvertent assumption that one's own site is "typical" and the tendency to place far more credence in what one has seen with one's own eyes than what is reported by another.[28] But as Hausfater went on to stress, the problem for those who wanted to discuss the biological and theoretical functions or causes of infanticide at the moment was that there simply wasn't enough information available. Very few people had seen infanticide, and in any case *seeing* infanticide was irrelevant: what was relevant was the development of long-term demographic and reproductive information on the length of male tenure, on female interbirth intervals, and on the number of males in a group—all of which would enable the sexual selection hypothesis to be tested mathematically. Rather than debating each other, Hrdy and Dolhinow would be better occupied in collecting these data. Pragmatically, however, this was impossible. By this stage Hrdy no longer had permission to work in India, and Dolhinow had largely retired from personal fieldwork. Instead she was supervising the only experimental colony of langurs in the United States as well as several graduate students who were continuing with fieldwork in Nepal. Neither woman would personally conduct prolonged primatological fieldwork in the future.

The Limits of Adaptation?

Working with one of her students, the next year Dolhinow published an extended response to Hrdy's work in *American Scientist*. Their account of this "heated controversy" (Curtin and Dolhinow 1978, 468) and of langur behavior was firmly based in the socioecological consensus that had emerged among primatological researchers in the late 1960s, discussed in chapter 2, where established variation in primate social behavior, both within and across species, could be accounted for by the particular ecological context in which a given primate group was found. From this perspective, langur infanticide was to be explained not as an adaptive reproductive strategy common to the primates, but as an unusual and pathological response to abnormal environmental conditions.

First, and in direct contrast to Hrdy's attempt to interpret the various accounts of langur social behavior as examples of a common underlying structure, they put forward a bifurcated model that stressed two distinct patterns of langur troop organization, based on the number of adult males present. From this perspective, the multimale troop pattern "probably occurs over the species' whole range and persists under all but the most stressful environmental conditions" (1978, 468); in sharp contrast, "the one male troop/male takeover pattern has been observed *only* where langur population densities are high and where human influence on the ecology is immense" (Curtin and Dolhinow 1978, 474; italics in original). Within multimale troops, while langur life was almost as peaceful as Jay had found it, further studies had revealed that male rivalry *was* an important element in troop life. The vital point, however, was that this rivalry was expressed and managed through various mechanisms ranging from simple threat behavior to the temporary exclusion of a male from the troop. These mechanisms make it possible for males to coexist, but they all "take time, and perhaps most importantly, they take *space*" (1978, 471; italics in original). This was the reason for the one-male pattern at Dharwar and elsewhere: where populations were crowded and stressed by human interference, there was simply no room for these evolved strategies to be expressed, and males could not coexist within the troop. Understandably then, in these specific circumstances "male rivalry may produce new and maladaptive results" (1978, 471) by triggering the periodic troop invasions seen at the infanticidal sites. In other words, those events did not represent one aspect of a general pattern of langur troop development but were a perversion of the normal multimale model of troop organization.[29]

Having directly contradicted Hrdy's central and fundamental claim that bisexual langur groups were normally characterized by the presence of a single male, the only male with the opportunity to breed, Curtin and Dolhinow went on to question the empirical basis of her account of infanticide in primates. They briefly reviewed the recorded events at each site to demonstrate that the "circumstances surrounding the disappearance of some 40 infant langurs" were not fully known (1978, 471). Hrdy had claimed that thirty-nine cases of infanticide were recorded in the literature, but at Dharwar only one attack by a male on an infant was directly observed, at Jodhpur only three had been seen, and at Hrdy's site, where the "events surrounding these dramatic levels of infant death remain somewhat obscure . . . [o]nly the reports of casual bystanders linked male langurs to any infant deaths" (472). Hrdy's "local informants" have now become "casual bystanders," and the status of their observations is discounted accordingly. Essentially, what is to count as an observation has been recast so as to exclude the testimony of nonprofessional observers,[30] and they point out that in most cases it was impossible to establish the identity of the infant killer. Information on impregnations and births following infant death or disappearance was similarly hard to come by. Equally unclear, they argue, was how a male might recognize his offspring or deduce a rival's paternity. There was, as Hausfater argued, simply not enough information even to test the hypothesis, much less recognize it as an evolved, adaptive strategy.

Alongside this trenchant critique of the infanticide database, Curtin and Dolhinow reordered the events of Hrdy's narrative, in particular challenging her presentation of cause and effect. In fact, they turned Hrdy's account back on itself, pointing out that an equally strong hypothesis could be made by *reversing* the model's connections. They accepted that there "is no question that males have definitely injured infants following some take-overs," but they did wonder "whether *infants* are the targets of attack." Rather than the infant's death triggering the mother's return to receptivity, they proposed that the mother's sexual readiness, brought about by the appearance of the new male, might just as likely lead to the infant's accidental death in the excitement surrounding mating. They concluded by asserting unequivocally that the sexual selection model "explains not the results of observation, but the products of assumption": assumptions about whether the infant was deliberately killed, about which animal killed the infant, and about what effect infant disappearance might have on the mother's reproductive system (473). In all of India, they emphasized, "there has been clear-cut evidence of only four langur infants being killed by males" (473), at two sites that have been shown to be ecologically and socially atypical. Infanticide, they concluded, is not a natural fact of langur life but an artifact of recent human pressure on the local environment.[31] It is a product of historical, not evolutionary, time, and rather than representing a regular pattern of primate reproductive behavior, it is an unfortunate consequence of serious social disorder.

Hrdy's response, as in her previous public interchange with Dolhinow, was to attack the uncritical use of "normal" (Hrdy 1978). She pointed out that since there are no objectively agreed-on criteria for the "normality" or "abnormality" of a site,[32] Curtin and Dolhinow were able to use their own somewhat circular definition: infanticidal sites must be abnormal because infant killing was seen there.[33] Noninfanticidal sites are normal because infanticides don't occur, even if they are found at the very limit of the species' ecological range—as in Nepal[34]—or if the langurs there were subject to serious harassment by humans. Just as Curtin and Dolhinow criticized her model of sexually selected infanticide for the number of assumptions it was forced to make in the absence of data, so she condemned their presumption that human harassment and population density are linked as cause and effect and that social stress is characteristic of densely packed populations. While population density may well influence levels of infanticide, she argued, it was at least possible that Dharwar had high population density because the forest/town edge was an ecological niche where langurs can flourish. But ultimately, for Hrdy, the fundamental element that opponents of the sexual selection hypothesis must account for was that infanticide had now been reported for so many species of primates under a wide range of ecological conditions. In their concentration on the details of the reported events at Dharwar, Jodhpur, and Abu, Curtin and Dolhinow did not address this question of the comparative data at all—a pattern that was to be repeated later in the debates.

This omission is important because of the way comparative work—both within and between species—can be seen to function for field research as a rhetorical and practical substitute for replication with laboratory studies, a practice that is as much rhetorical as it is "real." Curtin (1977) had already raised the role of experimentation in elucidating the causes and functions of infanticide when he called for integrating experimental work in both the field and the laboratory with long-term field observations (1977, 34), although he acknowledged the difficulty of conducting experiments in the field. Experiments related to infanticide carried the danger of serious damage, and even death, for infant and adult members of this protected—and in Hinduism—revered species. Angst and Thommen's review had made it clear that any work that risked the deliberate killing of infants was unethical (1977, 215).[35] The functional alternative for studying variation in social and ecological conditions that might produce infant killing was therefore comparative work. The question was the relevance of the comparative material: Was attention to be directed exclusively to the langur data, or were researchers to cast their nets more widely?

Specifications and Generalizations

Increasingly, this contrast between a tendency to draw on the general comparative literature and the urge to focus on the specific context of the infanticidal langur sites became a central factor in the dispute. Outside primatological circles, infanticide was becoming accepted as an adaptive reproductive strategy, one that had first been identified in langurs but was now increasingly recognized as widespread throughout the animal kingdom. This can easily be seen by comparing different editions of John Alcock's influential textbook *Animal Behavior:* whereas the first edition (1975) discussed the adaptive possibilities of infanticide only in relation to humans,[36] by 1979 the competition hypothesis was explained in detail and treated as the ideal type of mammalian infanticide. Although the "controversial case" of the langurs was described in the first chapter, it was only to show how different observers could produce widely divergent accounts of animal behavior: by 1979 the idea that infanticide probably functions as an adaptive part of primate social and genetic evolution had become part of the standard animal behavior curriculum. Certainly most of the primatologists who accepted this hypothesis seemed aware that unequivocal evidence, either for the infanticidal events themselves or for the wider social and reproductive context, was still lacking. But the potential genetic advantage that sociobiological theory predicted for males that kill unrelated infants seemed large enough to make repeated, unequivocal observations of the event somewhat irrelevant. In any case, unambiguous and complete observations of rare events were well known to be very hard to achieve under field conditions—for example, very few incidents of predation on primates had been observed in their entirety.

Certainly Hrdy's substantive review article in 1979 dealt with the existence and adaptive nature of infanticide as established fact. This paper drew on data from field studies of ground squirrels, hyenas, lions, fish, insects, primates, elephant seals, wild dogs, and eagles as well from laboratory work on rodents to demonstrate both that infanticide is widespread and that it "cannot be considered as an unitary phenomenon" (1979, 14). Animals carry out infanticide in a wide array of different circumstances, which she classified into four main categories: exploitation, resource competition, parental manipulation, and sexual selection. The important point here is her concern to outline the conditions under which one would *expect* these different categories of infanticide: she was emphasizing the conditions that would *predict* infanticidal events. Thus her discussion of a fifth and final class of infanticide—social pathology—is limited, since as a hypothesis, she argued, it generates no predictions but merely asserts that infants are the most vulnerable members of the group when aggressive social change occurs. For primates,[37] while she acknowledged that "much of the evidence on infanticide is based on isolated cases and on inferences derived from witnessed attacks which were followed by disappearances of infants" (1979, 20), she was emphatic that the observed conditions of most primate cases fitted the male sexual selection hypothesis. That is to say, an incident could be "counted" as an infanticide if the *pattern* of observed events fitted those predicted, even if the event was not witnessed throughout by an accredited observer.

Again, as in her earlier articles, Hrdy was eager to demonstrate the fruitfulness of the sexual selection hypothesis for stimulating further research, especially in species where infanticide had not been recorded—where its *absence* was something researchers must explain. But perhaps her widest claim for the productiveness of the hypothesis came at the end of her review when, having dealt with the specific questions and examples raised in the langur controversy, she turned again to the comparative literature—but this time, the literature on humans. She raised two interrelated and unanswered questions of human evolution. Why are human females continuously sexually receptive, even during pregnancy, and why do they not advertise their moment of peak fertility as do their close relatives, the chimpanzees? Both characteristics, Hrdy suggested, could well be the product of a situation in which "it was beneficial for females to confuse paternity, as in the case of infanticidal species" (1979, 34). By hiding ovulation and by possessing the capacity to mate with males regardless of her reproductive state, a female can render paternity profoundly uncertain and thus make infanticide an extremely high-risk strategy for males because of the danger of killing their own offspring. Hrdy's use of the comparative and anthropological literature thus enabled her to demonstrate the sheer power of her sexual selection model—it was not just an explanation of a rare behavior of a particular subpopulation of langurs but could also be used to elucidate some of the most vexed questions of human evolution.

In contrast, Jane Boggess, another of Dolhinow's graduate students, chose to concentrate her review of infant killing solely on the langur studies.[38] Like Curtin and Dolhinow before her, Boggess argued that the sexual selection hypothesis rested on a number of assumptions—those made in Hrdy's "type cases" of troop structure and infanticide—and reviewed the data from Dharwar, Jodhpur, and Abu in detail to show that few of these reported examples of infant killing bore out the predictions and assumptions of the sexual selection hypothesis.[39] Like Curtin and Dolhinow, she stressed that there were two clear models of langur social structure, although "structure" may not be the most appropriate word to use in this context, since Boggess emphatically denied that the events at Dharwar, Jodhpur, and Abu represented any kind of "pattern." She argued that the term "takeover" enabled Hrdy to gloss over great variety in the events the word was used to describe, since when recorded events were examined in detail it seemed clear that "different social phenomena were producing the same results: sudden and complete replacement in adult males and attendant infant mortality" (Boggess 1979, 87). Similarly, the aggression shown toward infants was extremely variable—consisting of attacks on mothers with infants, brief attacks, prolonged attacks, attacks by single males, attacks by groups of males, group hysteria, even group nonchalance—thus making it "unreasonable to explain infant killings by a single cause" (88). It might be, she suggested, that females themselves are provoking these attacks by exhibiting hostility to new males or by displaying to them.

Finally, and for the first time in print, Boggess outlined the nature of the data that opponents of the sexual selection hypothesis would require before accepting it as plausible. Since Hrdy's hypothesis depended on a correlation between infant killing, nonpaternity, and the subsequent fertilization of females, it must be shown that the male was not the father of the killed infant, that he fathered the next offspring of the infant-deprived mother, and that the interbirth interval was significantly decreased. In the absence of such data—which either did not exist or, she argued, did not support the sexual selection hypothesis—the hypothesis cannot be tested. Like Curtin and Dolhinow, she turned to socioecology as the ultimate explanatory factor behind infanticide. Twenty years of records of variability in behavior that could consistently be related to different environmental contexts have, she argued, been sidelined by the current fashion for "the tenets of genetic investment theory" (100). This predisposition had created a context in which "social crowding and artificially high density" can be considered typical rather than, as the socioecological perspective suggests, a situation in which "the species-typical characteristic of male social instability [is operating] against the reproductive success of all troop members including the new male residents" (104). Attacks on infants were maladaptive: by no stretch could they be regarded as normal elements of langur life.

Infanticide in Primate Life?

Boggess and, to a lesser extent, Hrdy were concentrating on the competitive evaluation of their preferred and mutually exclusive hypotheses. However, if we consider the rest of the primatological literature over this period (1977–82), we can see at least three distinct trends in relation to infanticide. First, and most generally, there was an evident drift away from evaluating behavior in terms of contributions made to group survival and toward a focus on the individual—or indeed selfish—consequences of particular behaviors. In 1978, for example, Craig Packer published an influential article that assessed the conditions under which group selection might be a factor in the evolution of primate society and concluded that in almost every case "predictions from individual selection are much more likely to be correct" (Packer 1978, 110). In another case, Scollay and DeBold's (1980) conclusion that "allomothering"[40] could not be adaptive because it could not possibly benefit both participants[41] was immediately contradicted by Sam Wasser and David Barash (1981), who pointed out that if allomothering was treated as a selfish rather than an altruistic behavior, its adaptive benefits were clear.[42] Similarly, Mohnot (1980) described intergroup infant kidnapping by females at Jodhpur as exploitation. Researchers seemed more and more willing to evaluate behavior so as to emphasize conflict between individuals rather than cooperation within the group.

However, although researchers increasingly accentuated the significance of selfish behavior by primates, and while the "most striking example of an individual increasing his own inclusive fitness at the expense of the group as a whole [was] that of infanticide" (Packer 1978, 107),[43] alternative explanations continued to be proposed. It was by no means true that the sexual selection and pathological hypotheses were the only possible solutions. Rudran (1979), now studying howler monkeys in Venezuela, where infanticide was the most frequent cause of death in his study population, argued that food competition and population regulation were at its heart. Using the model he had developed earlier in relation to the langur data

(Rudran 1973; and see chapter 3), he argued that infanticide frequencies would vary with population density, acting to regulate population in relation to the carrying capacity of the habitat. Suzanne Ripley (1980) also argued that infanticide operated as an essential means of population control, arising out of the need for langurs to maintain genetic polymorphism within the population and to control their numbers. The benefits to males are simply the proximate means for achieving these goals: "The adult male langur who eliminates his rival's offspring, for whatever motivation, in a situation of environmental flux, high reproductive rate and detrimental inbreeding, contributes positively, if unwittingly, to his own species viability" (1980, 363).

In a similar vein, Kawanaka (1981) suggested that infanticide in chimpanzees might best be understood as eradicating infants born of incestuous unions. While all these examples treated infanticide as adaptive, for the group if not the individual, at least one researcher tried to sustain a different version of the social pathology hypothesis. Rijksen (1981), having observed infant killing at the Noorder Zoo in the Netherlands, argued that the behavior was the by-product of males' competing with infants for females' attention, with disastrous consequences for the infants. In the wild, he argued, the same behavioral complex comes into play when males encounter strange females. Rather than being adaptive, infanticide is thus a side effect of a strategy for swift achievement of social recognition and dominance.

Even where researchers accepted the sexual selection hypothesis, some remained wary of what they saw as the uncritical adoption of Hrdy's model. Others were careful to emphasize the preliminary nature of their conclusions. S. C. Makwana, a researcher at Jodhpur, recorded several instances of social change and infanticide or infant disappearance[44] but was emphatic that "of the four changes of leadership observed, only one was followed by infanticides," which suggested that infanticide was "unusual and is not the common practice among primates" (1979, 299). Also at Jodhpur, Mohnot, Gadgil, and Makwana (1981) reviewed eleven years of population census data at the site and found no support for Hrdy's cyclical model of "initial/growth/mature" phases in the size and history of the langur troops there. Butynski's work on blue monkeys in the Kibale Forest, Uganda, found that males that eliminated a rival's offspring would gain an advantage, but whether they would also increase their own reproductive success remained moot. In this context, while infanticide could be treated as "part of a flexible, adaptive reproductive strategy of males" (1982, 18), data on male tenure length was essential for evaluating the advantage the male might attain.[45] This echoed a call that Chapman and Hausfater (1979) had made when they published their mathematical modeling of infanticide's impact on reproductive success. Having outlined their parameters and stated the conditions for the stability of infanticide as an evolutionary strategy, they found that the crucial "demographic and reproductive data are not available even for well-studied populations" (1979, 239). As a result, the thorough testing of their model would have to await the collection of such information.

The Absence of Data

But at the heart of the controversy lay fundamental disagreements about the nature of the data required to test the sexual selection hypothesis. In July 1980 a symposium was held at Pisa to discuss langur reproductive strategies (Hausfater and Vogel 1982),[46] and while none of the well-known American opponents of the sexual selection hypothesis were listed as present, the symposium's organizer, Christian Vogel, was vociferously hostile to the use of sociobiological concepts and hypotheses, including the adaptive account of infanticide (Vogel 1973, 1977).[47] This symposium had two main aims—to review the data currently available on langur social and reproductive behavior in the light of the ecological characteristics of the sites where they were studied, and to identify the nature of the data needed to test the various infanticide hypotheses, whether they dealt with reproductive strategies, population regulation, psychosocial factors, or resource competition.[48] However, it was evident that the controversy had now polarized to such an extent that reaching agreement on the kind of data needed to evaluate the competing hypotheses was very difficult.

The sexual selection hypothesis was at the root of the dissent. As Hausfater and Vogel acknowledged, "At one extreme in this debate are individuals who suggest that most reports of infant killing by adult male langurs are at best the product of incomplete observations subject to imaginative interpretation. At the other extreme are researchers for whom any harsh gesture by an adult male toward an infant is taken as evidence of the actual or potential infanticidal proclivities of that male" (Hausfater and Vogel 1982, 163). In an attempt to establish common ground where the two extremes might meet, they outlined what they considered to be the basic standards for reporting infanticidal events. They suggested that the word "infanticide" had been used far too loosely in the literature, covering everything from fatal biting to aggressive threat, and that in many cases there were serious grounds to doubt that the new male was in fact responsible for infant disappearances.[49] To resolve this, they argued, those who reported alleged infanticides had a greater responsibility than usual to include every possible detail in their accounts. These particulars, reminiscent of Carpenter's description of the ideal field report, covered everything from the timing of the male's entry into the group to the precise details of the delay between infant wounding and disappearance, followed by reproductive data on the mother from the moment of infant loss to the next estrus. However, this demand for detail should not, they suggested, be taken to extremes. Arguing by analogy, they pointed out that

> if a leopard were observed to jump into the middle of a monkey group and then to flee with adult males in pursuit, holding a female in his mouth, few observers would seriously question that an act of predation had occurred, particularly so when later the adult males, but not the female, were observed returning to the group. Yet in analogous descriptions of infanticide, it has been seriously suggested that the mother may have killed her own infant in the confusion surrounding her struggle with an adult male attacker. (Hausfater and Vogel 1982, 164)

Reports of infanticide must be detailed, but unrealistic standards—when compared with other rare and swift events such as predation—should not be imposed.

Hausfater and Vogel identified three other problems with the way the infanticide database was treated. In several cases participants in the debate had tried to tabulate the number and frequency of infanticidal events at particular sites as a means of evaluating the importance or innateness of the phenomenon. While it was true that infanticide reports had been made from only a limited subset of the total range of langur study sites, it was also true, they stressed, that these sites themselves might not be a representative sample of the langur population.[50] Additionally, infanticide was rare, and raw numbers meant little in themselves: again, predation on primates was observed very infrequently, but there was little doubt that predation not only occurred but was important in the evolution of primate populations. Finally, the authors pointed out that there was a consistent assumption in the infanticide literature that either infanticide occurred at certain sites (in which case all local males should be infanticidal) or it did not (in which case no males should be infanticidal). However, there was no reason to make this assumption: it was possible, as Chapman and Hausfater's (1979) computer model had predicted, that infanticide could exist as a behavioral polymorphism, where a certain *proportion* of males might be infanticidal. Alternatively, it might be one of a range of male reproductive strategies, where infanticide was an option for males in certain circumstances but not others.

Having offered these caveats, Hausfater and Vogel outlined and evaluated the four main causes underlying the hypotheses offered at the symposium to explain primate infanticide. Their conclusion, not surprisingly, was that it is "premature to rule out nearly any of the above hypotheses concerning infanticide in langurs" (1982, 170). No decisions could be made until more basic demographic and reproductive data from prolonged field studies were available—details of average male tenure lengths, female interbirth intervals, and relative levels of population density in different ecological contexts. However, "shortly after this symposium, the Government of India decided not to issue extensions for the long-term studies of Indo-German and American research teams" (171). Without research visas allowing them to continue their fieldwork, most langur sites were now closed to researchers, although work on other species and on captive langur colonies could continue. However, since those who opposed the sexual selection hypothesis focused overwhelmingly on the data from field studies of langurs, such alternative sources of information were unlikely to be acceptable in resolving the controversy.

Sociobiology, the Social Sciences, and Primate Fieldwork

By the early 1980s it was evident that interest in the debate was no longer confined to the natural sciences but had spread to the social sciences as well—perhaps inevitably, given the significant links between infanticide and sociobiology. Glenn Schubert, a professor of political science at the University of Hawaii,[51] published an article in *Social Science Information* (a journal aimed primarily at sociologists and anthropologists), that not only argued that sociobiology had become a scientific religion complete with "new dogma [and] demi-godlike leaders" (Schubert 1982, 199), but singled out the sexual selection hypothesis as an example of the "stretching of the evidence" that sociobiology is based on (200). In particular he sought to demonstrate that "sociobiologists ought to be required to play the game of field research by the same rules of normal science that restrain the rest of us" (201). This is a key point: although he castigated Hrdy for failing to recognize that her interpretation of events at Abu was colored by ideology at least as much as Dolhinow's work was influenced by Radcliffe-Brown's anthropology, he concentrated on what he considered the unscientific *consequences* of this lack of self-analysis. In particular, he was immensely critical of what he saw as the inappropriate relation between sociobiological theory and empirical data in Hrdy's work.

Like every other critic, Schubert reviewed the reported events at Dharwar, Jodhpur, and Abu to demonstrate how they failed to conform to Hrdy's proposed model as well as stressing the scarcity of witnessed infant deaths. However, he went much further, not only criticizing Hrdy for depending on observations by amateurs, but also seriously questioning her own status as an authoritative observer. In particular, he was opposed to Hrdy's decision to include in her analysis the observations made during her first field season: arguing that this period of observation was equivalent to the "pretest" required in social science methodology, he pointed out that no "competent social scientist" would include such a preliminary survey in final results (Schubert 1982, 212).[52] But while the paucity of the records themselves was a problem, it was Hrdy's attitude to the database that pained him—she did "not purport to present a theory derived by induction from analysis of her data" (219), he argued, but instead was presenting only the evidence that supported a hypothesis derived from neo-Darwinian principles. In contrast, "All of the great discoveries of classical ethology, by Lorenz and by Tinbergen and by their students, came from persons who developed hypotheses about animal behaviour in an endeavour to make sense out of data they already had in hand—not from them going around looking for data that would make some sense out of their theories" (219–20). While Schubert also raised a range of other problems (lack of reliably identified individual langurs, dependence on local informants, the definition of a takeover), he reserved his most stringent attacks for Hrdy's approach to doing science in the field, arguing that it did not correspond to the correct use of the scientific method.

The editorial board of *Social Science Information* invited Hrdy to respond to Schubert's critique,[53] and her response concentrated on Schubert's misunderstanding of the nature of data collection in the field.[54] Along with other fieldworkers, she would "be the first to admit that [field] reports suffer from small sample sizes, gaps in data collection and the absence of information on lifetime, or even long-term, reproductive success of individual monkeys" (Hrdy 1982, 245), but such lacunae were the inevitable consequences of studying free-living animals. Schubert's attack on her work was, she suggested, an example of a more general conflict between "the philosophy of science and the practice of fieldwork with wild animals" (247), since the falsification of hypotheses so beloved of positivist philosophy is virtually impossible to achieve in the field. Returning to the question of experiment, she pointed out that while this strategy was undoubtedly central to disproving hypotheses, conducting experiments on infanticide would involve the capture and death of adult and infant langurs (since for opponents of the sexual selection hypothesis, only a dead body can "count" as an example of infanticide) in large numbers (since small sample sizes are inadequate). Not only would such activities be ethically undesirable and illegal, they would have drastic political consequences: "If a foreigner attempted to conduct such experiments in India (where Americans are unpopular and langurs sacred) the responses would range from local riots to an international incident" (1982, 248).[55] In place of langur experiments, Hrdy again turned to the comparative evidence for infanticide—both for other primates and for rodents, where experiment was possible and had produced evidence strongly supporting the sexual selection hypothesis. But the wider point remained—philosophy of science was not developed to understand the processes involved in science in the field and did not take its constraints into account.

But in many ways it must be admitted that sociobiological theory is equally ill suited for testing in the field. In a review of the impact sociobiology has had on primate field studies, Richard and Schulman (1982) argued that while the perspective has made a major contribution—filling a theoretical vacuum, casting new light on old and intractable problems of primate behavior—at the same time it has been accompanied by very real problems. To test most sociobiological hypotheses, for example, data were required on the long-term demography and behavior of large samples of known individual primates as well as long-term information on resource distribution. Additionally, knowledge of paternity was central to evaluating sociobiological arguments, and some means must be found both of determining the cost/benefit associated with particular behaviors and of determining how animals might base judgments on these estimates. Finally, they acknowledged that there was an almost "total ignorance of the genetics of nonhuman primate behavior" and that many sociobiological models are of an "untestable nature, even under the best of circumstances" (1982, 244). Not all these problems are associated purely with sociobiology—although all had been raised in relation to the evaluation of the infanticide controversy—but they are questions, Richard and Schulman argued, that must be faced and ultimately solved by those who wish to apply and test sociobiological theories and concepts as part of their field research. Clearly, some were solvable, given enough time and effort, but others required judgment calls or close cooperation between field and laboratory researchers. In any case, it is clear from their conclusion that, with Hrdy, they accepted that biological hypotheses "are not amenable to test as hypotheses in the physical sciences can be tested. Rather, models must be evaluated and judged more or less plausible on a variety of grounds" (1982, 248). The precise nature of those grounds, however, could be identified only through cooperation between researchers—and in the case of infanticide, as this chapter has shown, such cooperation had become steadily more difficult to achieve.

The Wenner-Gren Conference

The extent of the dissent in which the debates surrounding primate infanticide were still mired can be seen in the first international conference on infanticide, held in 1982 at Cornell University in Ithaca, New York. Present were the key primatological supporters of the adaptive model of infanticide, but absent again were some of the critics of the sexual selection hypothesis—such as Dolhinow—who were represented by Jane Boggess. There were two notable points about this conference and the edited collection of conference proceedings that was published in 1984. First, the enthusiasm with which the various adaptive models of infanticide, including sexual selection, had been adopted by behavioral biologists working on a huge variety of species was evident. Second, and in contrast, the breadth and depth of the divisions between primatologists on the topic of sexually selected infanticide seemed immense, as Glenn Hausfater noted in his introduction to the primate section. He had to explain to the audience that the chapters were chosen both to provide a sample of the nonhuman primate database and to demonstrate the way different researchers had chosen to interpret the data. These conflicts, he suggested, might be confusing to those unaware of the "controversy" the subject had attracted within primatological circles (Hausfater 1984, 145). The extent of that controversy was illustrated by his description of the inability of three researchers—Boggess, Sugiyama, and Hrdy—to reach agreement on the langur data, necessitating the publication of all three papers.[56] Clearly, even though adaptive infanticide was presented as a straightforward research area for those working with other animals, for primates, and specifically for langurs, there was no imminent end to the controversy.

What is striking, however, is how far these three papers remain embrangled in the precise questions that had been raised from the controversy's outset. Boggess (1984) again returned to the specific circumstances of each claimed langur infanticide in an effort to demonstrate that in the rare cases where they were observed in anything close to their entirety, few (if any) of these alleged attacks fit the parameters of the sexual selection model precisely, and to show that the infanticidal sites were atypical as a result of human intervention in the local environment. Again, she stressed that the assertion that males gain reproductive advantage from infant killing remains an assumption in the case of langurs, since the data needed to test the hypothesis—from long-term, continuous studies—did not exist. In response, and while concentrating on the proximate causes of infanticide, Sugiyama (1984) challenged her presentation of Dharwar as ecologically abnormal, while Hrdy concentrated on Boggess's attitude toward the infanticide database. She admitted that the data were not of the best quality—an inevitable consequence of the constraints of field research—but argued that Boggess was imposing a double standard, where the "criteria for accepting negative evidence appears to be far less stringent than the criteria applied to data that support the sexual selection model" (Hrdy 1984, 317). It would not, she pointed out with heavy irony, "have occurred to [her] a few years ago to make the distinction Boggess does between killing an infant and inflicting life-threatening wounds" (319). But more important in accounting for the persistence of the controversy, she stressed again the significance of the consistency in patterns showed by the disparate sources of data rather than the preciseness of the fit with the model—especially, again, in relation to the comparative evidence.

And it was clear by this stage that the comparative evidence for infanticide as an adaptive strategy was mounting. The collected conference proceedings included work on mammals, amphibians, reptiles, insects, and birds. For other primate species, Rana Sekulic (1983) had published work on howler monkeys suggesting that infanticide occurred even where changes of male membership were gradual rather than the dramatic changeovers seen for the langurs. In another study population of howlers, Clarke (1983) found that the infants of high-ranking females—the ones most likely to have been sired by the former leader—were targeted by the new dominant. Packer and Pusey's work on lions in particular was demonstrating that clear female counterstrategies appeared to be at work against infanticidal males (1983). As I mentioned above, an entire section of the Wenner-Gren conference was devoted to infanticide in rodents—species where experimentation was permissible. It is perhaps not surprising that Hausfater and Hrdy felt optimistic enough in their introduction to speculate that

> quite possibly, readers ten years from now may take for granted the occurrence of infanticide in various animal species, and may even be unaware of the controversies and occasionally heated debate that have marked the last

> decade of research on this topic. Such readers will thus most likely be puzzled by the obsessive reiteration in this volume that infanticide is a natural and not necessarily pathological behaviour. (1984, xi)

Conclusion

Clearly, by 1984 there still existed a range of hypotheses to account for infant killing in human and nonhuman animals, but all except one (social pathology) treated infanticide as an adaptive event. Sexual selection was not the only explanation—in certain circumstances it could be expected to occur as a result of parental manipulation or of resource competition, for example—but it was the explanation cited most consistently for primates, and it was the account that continued to attract hostile criticism. Explicitly, the sexual selection hypothesis was dependent on an evolutionary account of behavior that emphasized competition between individuals and that required accepting a genetic component to behavior that could not be unequivocally demonstrated. However, by this stage most primate researchers had accepted the hypothesis. Those who continued to oppose it publicly were closely associated with Phyllis Dolhinow—who had herself withdrawn from direct involvement in the debate, although she continued to support the social pathology explanation—and appeared to be increasingly isolated and beleaguered.

Understandable strategies had been adopted by both sides in the dispute over primate infanticide. On the one hand, those who supported the hypothesis emphasized the existence of clear patterns, even if those patterns could not reliably be seen in their entirety, and turned to the steadily increasing array of evidence from the comparative literature to bolster the frequently fragmentary reports from the primates. On the other hand, those who opposed the hypothesis found such an attitude toward the evidence slipshod, stressing instead the need to consider each proposed example of sexually selected infanticide as a whole, to compare it with the projected pattern of the sexual selection model, and to reject it if it did not measure up. For this group of researchers the controversy was really about the behavior of some langurs at sites in southern India, and how far it was justifiable to use their behavior to generalize about other primates: first, it was necessary to see whether the initial step in the logical chain Hrdy constructed was acceptable—and from their point of view, in all but a few of the cases it was not. Broadly, as chapters 6 and 7 will show, these two perspectives can be characterized as the sociobiological and the socioecological. The sociobiological revolution spread more slowly in primatology than it did for other examples of animal behavior, and the proponents of the older, ecological paradigm were far more vocal in defending their position. At the same time, however, the sheer volume of knowledge about primate behavior was expanding exponentially, and simply adding to the recorded literature was becoming insufficient: behavior was not simply to be recorded but also needed to be explained, and the concepts of sociobiology seemed to provide a more direct and targeted way of doing so than did those of pure socioecology.

It seemed impossible to reach agreement on how to resolve the controversy. However, the theoretical divisions were exacerbated by the conditions of fieldwork itself. Where free-living animals inhabit forests and savannas that they are far more physically suited to exploit than is the awkward human tracking them, it will be impossible to observe all interactions in their entirety. Rare and rapid events, such as infanticide, while spectacular, would be even harder to record accurately and in detail. In addition, certain kinds of information were almost impossible to acquire in the field, especially at this stage of primatological research. A key requirement for evaluating the infanticide hypothesis was obviously infant paternity and male lifetime reproductive success—but paternity is notoriously hard to evaluate, and male monkeys tend to transfer between troops, whereas primatologists tend to focus on particular troops. Experiment could theoretically have been used to judge the competing hypotheses, but it was a practical and ethical impossibility in the field. And access to the field sites itself depended on getting permission to carry out research: where this was denied, gathering any kind of data in the hope of reaching a conclusion was impossible. All these conditions also pertained to other aspects of primatological fieldwork—but in the case of infanticide or infant killing these taken-for-granted assumptions about the limits of what was practically possible were being directly challenged.

Despite these problems, the adaptive approach continued to gain strength and supporters. By the mid-1980s, many researchers considered the presumption that infanticide was the product of social pathology outdated and old-fashioned. The infanticide controversy, however, was by no means resolved, as the reviews of Hausfater and Hrdy's (1984) edited collection of the Cornell conference proceedings were shortly to demonstrate.

FIVE

Controversy Resurgent

Introduction

Neil Chalmers, who reviewed the edited collection of papers from the Cornell conference for the *International Journal of Primatology,* was open about his confusion in the face of the controversy. Having read chapters espousing the adaptive hypothesis, he was convinced: after reading Boggess's critique of the database, his views were seriously shaken. In the end, he concluded that "one is left not knowing what to believe" (Chalmers 1986, 328), since each author was capable of eloquently demolishing the case so carefully marshaled by another, and independent standards for assessing the evidence simply weren't available. While he granted that it was stimulating to see these fundamental questions about the nature of observation and the relation between theoretical backgrounds and the gathering empirical data so thoroughly explored, the controversy itself remained as impassioned and as intractable as ever.

However, the next few years saw the apparently inexorable rise to dominance of the adaptive interpretation of primate infanticide.[1] Publication after publication reported infant killings or infant attacks occurring under conditions that to a greater or lesser extent met the basic requirements of the sexual selection model: that the male not be the father of the victim and that he mate with the mother sooner than he would otherwise have done (Hausfater and Hrdy 1984, xix).[2] Jodhpur, the Indian site where infanticide had first been witnessed (Mohnot 1971) and then dismissed (Vogel 1977), was fast becoming the axis on which the adaptive model turned. The study there could now be characterized as long term, with researchers recording the behavior of recognizable individuals, behavior that could be considered in relation to the known life histories of the animals and their relatives and that regularly continued to include infanticide. Besides the reports on other primates, records from other animal species—in particular the long-term lion study taking place in the Serengeti, headed by Anne Pusey and Craig Packer—continued to support the argument that infanticide operated as part of a general male reproductive strategy that would be expected under given social conditions. To most eyes the controversy had reached closure by the early 1990s. Infanticide, although horrific and disturbing, was an adaptive and predictable response to certain circumstances.

However, in the early 1990s two events showed there was no immediate prospect that the infanticide controversy would rest in peace. The first was the conference held at Erice in Sicily, the second of the infanticide conferences to make its way into book form (Parmigiani and Vom Saal 1994). In this case the potential political implications that infanticide had for anthropologists seriously hindered the organization of both the conference and the book. The second event was a professor's suggestion to one of his graduate students that the infanticide literature would be a good subject to review for a term paper. This paper was eventually published in the *American Anthropologist,* reopening the primate controversy, to the shock of many primatologists. The increasingly acrimonious debate that followed also demonstrated that sociobiological theory was still a major problem for primatologists in particular, not least because of their close connection with—and often, institutional location within—anthropology departments. This chapter examines the stabilizing consensus that had formed by this point and the nature of the new critique. It shows how the controversy, often ascribed to political or nonscientific motivations by participants in the debate, was in fact the product of at least two distinct approaches to doing fieldwork in primatology and the difficulties of interpreting and reconciling the central, theoretically distinguishable but practically inseparable, influences over primate social behavior: ecology and biology.

The Stabilizing Consensus

Data from langur studies, from other primates, and from other animal species including humans formed the nucleus around which the adaptive consensus was coalescing by the early 1990s. Infanticides, of course, were found to occur in many circumstances, and infanticide in other animals did not always take the form of reproductive competition, the model that had provoked most controversy. However, there were enough examples of sexually selected infanticide in various taxa to provide what supporters of the adaptive model considered excellent comparative evidence that infanticide in primates was a result of sexual selection. This model, especially in relation to male-male competition, was the most common context in which infanticide was seen for primates—and the field site at Jodhpur, near Rajasthan in northwestern India, provided the most numerous and most detailed reports of infanticidal attacks by males.

Jodhpur

As chapter 3 described, S. M. Mohnot of Jodhpur University had initiated work at this site in July 1967, under the supervision of M. L. Roonwal and concentrating on three troops: B25, B26, and M27. Other researchers from the university continued work there through the sixties and into the seventies (Makwana 1979; Mohnot 1980; Mohnot, Gadgil, and Makwana 1981), and by 1977 they were joined by scientists from the University of Göttingen in Germany. Christian Vogel and his coworkers maintained a collaborative presence at the site from 1977 until December 1982 (Vogel and Loch 1984; Winkler, Loch, and Vogel 1984), when the Indian government withdrew permission for foreign researchers to work in the country. The Indian members of the team then continued observations at the site until the Germans were able to return in the mid-1980s. This long-term study concentrated on three focal troops: the two bisexual troops first studied by Mohnot, now renamed Bijolai (B26, where the first observed infanticide had occurred), and Kailana-I (B25), which had split in two in 1977, giving the third focal troop, Kailana-II (Winkler, Loch, and Vogel 1984).[3] The members of these troops were individually known to the researchers, as were many members of the local male bands, providing an unrivaled opportunity to study Hanuman langurs not as members of age-sex classes but as individuals with unique life histories and to place those histories in the context of recorded site ecology. The only drawback was that the langurs were regularly fed by local people, who considered them sacred: however, as with Hrdy's work at Abu, this also meant that the animals had little fear of humans and thus could be habituated more effectively.

As a result, Jodhpur was an excellent site for beginning to test sociobiological hypotheses, hypotheses that normally require long-term data on the reproductive success of known individuals. Ironically however, Christian Vogel was known to be a vociferous opponent of sociobiological thinking. In his short introduction to an issue of *Human Evolution* devoted to the ongoing research at Jodhpur (Vogel 1988), he described his original assumption that "social systems have to be considered as a kind of supra-individual organism, i.e., higher-level self-regulatory systems, which per se form the central units of natural selection, and hence should be directly adapted in size, shape and structure to various ecological conditions" (219). From this perspective, individual langurs were functionally equivalent to the body's cells, and "adaptive" behavior meant behavior that "benefits the entire social system, its survival or maintenance, and its perpetuation" (1988, 219). As chapter 4 indicated, Vogel did not accept that infanticide could be considered adaptive. Unsurprisingly, the Göttingen group followed their leader's position, and Jodhpur, from being a key element in the case for adaptive infanticide, moved to the opposition camp.

However, this changed abruptly when Volker Sommer, a student of Vogel's, with his own eyes saw a male attacking and killing an infant. As Carola Borries, another of Vogel's students,[4] describes it, when he heard the news, Vogel was preparing his contribution to Hausfater and Hrdy's edited collection of papers from the Cornell conference. At that point he and Helmut Loch had

> pulled together all the evidence that's out there from Hanuman langurs to prove that it's just not really important, it doesn't really happen, if you really look deeply, there's just one or two cases that have actually been seen and not these twenty or something cases, so they were not really convinced that there will be infanticide, and then they got a telegram from Volker Sommer telling, I've seen it, and I took photos of it! And then the whole institute turned, and also my adviser [Vogel] changed his mind and changed his teaching and really got into it.

For this reason, while Vogel and Loch's chapter emphasized the weakness of the empirical database for infanticide and the need for further study and more details, it also acknowledged the significance of Sommer's 1982 observations. Vogel himself in 1988 described how events at Jodhpur had forced him to abandon the idea that group selection had any influence on how animals behaved: so many of the observations at the site made sense only in the light of individual decision making and the consequences of those decisions for the individual's reproductive success.

Certainly, the special issue of *Human Evolution* that showcased the Jodhpur research reflected their focus on individual reproductive success and sociobiological theory, as Borries's work on grandmaternal behavior (1988) and Rajpurohit and Mohnot's account of the fate of ousted male residents (1988) demonstrated. But it was the infanticide work that seemed particularly impressive. In 1987 Sommer had published his analysis of the infanticide controversy in light of the new observations at Jodhpur. Having acknowledged the "considerable disagreement" (163) that characterized the discussion of the function of langur infanticide, he accepted that the paucity of directly witnessed infanticides was a crucial problem with the debate so far and that Hrdy's conclusion in 1977 that infanticide was a regular and widespread occurrence had been premature. The critics had been correct to call for more data on the circumstances of infanticidal attacks and more information on infanticide's impact on reproductive success. In this paper, however, Sommer asserted that he could now present "detailed qualitative descriptions along with *photographic documentation* concerning six eyewitness and five suspected cases of infant killing during three infanticidal male replacements" (Sommer 1987, 164; emphasis added). Not only had these infant deaths been witnessed, they had been visually recorded and could be evaluated against both the long-term records at the Jodhpur field site and reports of infanticide made elsewhere.

Sommer outlined the various hypotheses that had been proposed to explain primate infanticide and rejected all but the sexual selection model. Social pathology, he argued, depended on the presumption that the animals are crowded and subject to human influence: Sommer noted that infanticide had apparently not been seen at the Berkeley colony (run by Dolhinow), even though the animals lived under captive conditions, and that Newton had recorded infanticide at the undisturbed Kanha site (Newton 1986, 1988). Population regulation and incest avoidance both required the operation of group selection, for which there was no known mechanism, and did not explain why only certain classes of animals were attacked—neither did the suggestion that males were motivated purely by proximate causes of sexual frustration or aggression. Similarly, resource competition left open why only males killed and why animals more vulnerable to attack (juveniles, for example) were left unharmed. Neither of the hypotheses that were directly concerned with the proximate social conditions of male replacement was borne out: rather than enabling the males to assert their dominance or to speed up their bonding with the females, infanticide actually prolonged the agitation. The only explanation that generated testable predictions and also explained events was sexual selection, which focused on the reproductive advantage the individual accrued by infanticide. In fact, it also generated at least two further hypotheses, one of which—the suggestion that females should experience induced abortions in the presence of strange males—seemed to be upheld when Agoramoorthy et al. (1988) produced their analysis of miscarriages at Jodhpur over a ten-year period. Seven were recorded in total, and five of these occurred during male replacements. In summary, while the data from Jodhpur did not fit *exactly* with the predictions generated by the reproductive advantage hypothesis (Sommer 1987, 193), they still supported that model far more emphatically than any of the others.[5]

As the long-term study at Jodhpur continued, more examples of infanticides and abortions were added to the database. Sommer produced another review of the sexual selection hypothesis in relation to the Jodhpur results as his contribution to the Pisa conference in 1990. Again, the data were broadly in line with the predictions generated by the hypothesis, but admittedly there were some odd results. For example, a significant proportion of infants that one would expect to have been attacked by new males were not (Sommer 1994, 162).[6] Despite this, the data from Jodhpur still matched most closely with the sexual selection model: males were not killing their own infants, females became sexually receptive sooner than they otherwise might have, and in the great majority of cases, the infanticidal male had sexual access to the mother. Who fathered the subsequent offspring was slightly less certain, since males other than the perpetrator also had access, but in over 70 percent of cases it was "likely" that the father of the new infant had killed the mother's previous offspring (167).

In his discussion of the results, Sommer was refreshingly open about the relation between observation and interpretation, pointing out that a major source for the infanticide controversy could be found in the "self-fulfilling prophecy," since "if a langur researcher witnessed infanticide, even weak evidence for its occurrence at other sites is usually readily accepted. If infanticide was not seen with one's own eyes, one is likely to play down or even deny its existence for other sites" (Sommer 1994, 171). His own experience supported this: having been taught by Christian Vogel that sexually selected infanticide was a deeply flawed hypothesis, he neither wanted nor expected to see infanticide at the site. When an infant was attacked in 1980 but survived, the report he made at that time emphasized the infant's survival, not the attack. However, once he had seen infanticide occur, it became far easier and more reasonable to treat that 1980 attack retrospectively as an attempted infanticide. Echoing Hrdy's point that observation is affected by expectation, he suggested that some researchers might be so deeply influenced by their expectations that no evidence would "convert" them. On the other hand, converts like himself were also in danger of overinterpreting their observations, since "a protestant converted to catholicism is likely to be more papal than the pope" (Sommer 1994, 171).

Primate Studies, Sociobiology, and Infanticide

For langurs, then, long-term records of infanticidal attacks now existed for at least one site and had also been reported for at least one other langur study site where the population was undisturbed by humans (Newton 1986, 1988). In reporting these attacks, Newton pointed out that while it was impossible to falsify the sexual selection hypothesis in the wild (for ethical and practical reasons), it *was* possible to falsify the proposition that infanticide was the result of social pathology if animals were seen to commit infanticide at a low population density—as at Kanha Tiger Reserve, his study site. From his records, it was not the density of the population that triggered infanticide but its composition: where animals lived predominantly in one-male groups, competition between males was much more intensive and more likely to be expressed through infant killing. More generally, report after report of langur behavior that either mirrored or echoed the social structure and cyclical troop model first described by Hrdy (Davies 1987; Reena and Ram 1991; Van Schaik, Assink, and Salafsy 1992) had now appeared, and infanticide was becoming a key theme in accounting for primate social structure and biological development.

Where infanticide itself was directly reported, it was overwhelmingly discussed in relation to how well the events matched the predictions of the sexual selection model. Boer and Sommer (1992) described infanticidal events among monkeys at the Hanover zoo in terms that specifically excluded the social pathology hypothesis. Moos, Rock, and Salzert (1985) saw a male gelada baboon attack unrelated infants in a captive group where aggression was rarely exhibited. Butynski (1990) directly compared the behavior of blue monkeys in both high- and low-density populations and showed that infanticide was *more* likely in the low-density group.[7] Struhsaker and Leland (1985) described the behavior of an infanticidal male red colobus monkey—unusual since in this species males do not tend to transfer between groups—and pointed out that this animal was especially unusual in that he was conceived before his mother joined the group. The only animal he was related to, therefore, was his mother—and tellingly, her infant was the only one he did not attack when the aggression began. Tarara (1987) recounted the way a male baboon also began attacking infants within his natal troop, but in this instance the animal had only recently become the troop's dominant male: a trigger for infanticide therefore might be a sudden rise in status. Andelman (1987) suggested that the threat of infanticide might have contributed to the evolution of concealed ovulation in vervet monkeys, making it much harder for males to assess paternity, while Watts (1989, 1991) argued that the possibility of infanticide was a key reason behind the evolution of gorilla sociality itself. Gorilla females, it seemed, need males to protect their offspring from other males (see also Stewart and Harcourt 1987).

Overall, the sexual selection hypothesis—along with sociobiological thinking in general—had clearly become far more acceptable within primatology by the middle to late eighties. Sociobiology had replaced the earlier version of socioecology as the dominant perspective from which to analyze primate behavior: analyzing social roles had been replaced by assessing reproductive success, so much so that ecological influences were in danger of being overshadowed. Two influential textbooks that appeared at this time admirably illustrate the shift. Dunbar's 1988 analysis of primate social systems warned of the danger inherent in ignoring the fact that the expression of an animal's genetic heritage was constrained by the particular ecological and demographic conditions in which it was found: sociobiology and socioecology should be treated as complementary perspectives rather than as opposed. The example of infanticide showed that one had to include both orientations to fully understand behavior: there was no need to insist that infanticide must always occur in populations where it had once been seen if one understood that the expression of this inherited trait depended on particular socioecological conditions. Similarly, the collection edited by Smuts et al. (1987) emphasized the importance of socioecology by devoting an entire section to it. They noted ruefully that in their opinion the early work of Crook and Gartlan (1966) had failed to inspire primatologists to take up the challenge of elucidating the relation between social behavior and ecological conditions. However, the final chapter of this section of their book warned against treating behavior as the sole product of ecology: in several species, especially those where females transfer from group to group, the need for protection from potentially violent stranger males might well be one of the key factors in the evolution of group living (Wrangham 1987, 293–95; see also in the same volume Cheney and Wrangham 1987, 233; Smuts 1987a, 1987b).[8] Finally, and building on Chapman and Hausfater's earlier mathematical model that would account for the spread of an infanticidal trait throughout the population, there were further attempts to model the strength of the selective pressure for infanticide (Breden and Hausfater 1990; Yamamura et al. 1990; Glass, Hold, and Slade 1995).

In fact, as I described in the introduction, Van Schaik and Dunbar (1990) explicitly suggested that infanticide might be one of the fundamental pressures behind the evolution of permanent primate social groups. In

1990 they published a paper in *Behaviour* that took issue with the prevailing wisdom that monogamy in primates arose in situations where the infant's survival depended on support from both parents, not least because there are examples of monogamous primates (such as gibbons) in which little direct paternal care had been seen. Their paper tested several hypotheses for the origin of monogamy in these species against the primatological literature and concluded that the most likely explanation was the need for males to defend dependent infants from stranger males. They suggested in this paper that infanticide might, in fact, be "a 'hidden' force" (1990, 53), so that even where attacks on infants had never been seen—as with gibbons—it still might have imposed a major selective pressure. Even more daringly, they argue in their conclusion that this threat of infanticide may well be the reason for primate sociality more generally. Primate females' "vulnerability to infanticide has made it indispensable for them to have permanent bonds with others" (54). While other females are a possible source of support, males, which are larger and share an interest in the survival of their own offspring, are to be preferred. This, for the authors, "may explain why permanent male-female associations and some form of positive interaction between adult males and infants . . . are so widespread among primates in comparison to other mammalian taxa" (54). Illustrating the impact of this thesis, Charles Janson explained that while initially his response to the reports of infanticide had been

> Oh, those wicked langurs, look at what they're doing, but my nice paternal cebus wouldn't do that, you know. . . . It seemed very hard for me to believe that my study animal would do something like that because the males spend most of their time being very cooperative and helpful toward their offspring, defending the young of the group from aggression from other males, and so on and so forth, and I'd never really stopped to think about, well, if you are protecting them from other males, why are the other males being aggressive in the first place?

The only fly in the sexual selection ointment seemed to be humanity's closest relatives: the chimpanzees. Chimp infanticide had been recorded relatively frequently for East African populations, but not in circumstances that readily matched the predictions of the sexual selection model. Researchers had sorted these events into three categories: male intergroup infanticide, male intragroup infanticide, and female intragroup infanticide (Kawanaka 1981). The last category covered the strange behavior of Pom and Passion, a mother-and-daughter pair from Gombe that had killed and eaten other females' infants. These events might be explained by the resource competition model, since the daughter was eventually able to establish herself in territory that might have been occupied by the cannibalized infants, or simply by the desire to eat meat—either constituted an adaptive account of the behavior. In the case of the male infanticides, the situation seemed much stranger. Intergroup infanticide presented no problems to the sexual selection model, since the males were killing unrelated infants, even though they did not then have the opportunity to mate with their mothers. Intragroup infanticide in this context, however, meant that the males were almost certainly killing their own offspring. Initially Takahata (1985, 162) proposed that the males "might have been under the false impression that the infant had been sired by the male of another group," since the infant-deprived females were recent immigrants to the chimp community and tended to be found on the very edge of the male range (Nishida and Kawanaka 1985; Hiraiwa-Hasegawa 1987). Later Hamai et al. (1992) suggested that the infanticide and cannibalism were a means for Mahale males to control the females: the mothers of the victims had tended to mate preferentially with adolescents and subordinates rather than the dominants that attacked the infants. Whatever the explanation, at this stage researchers were forced to conclude that "in the case of chimpanzees, the adaptive significance of infanticide is still a puzzle" (Hiraiwa-Hasegawa and Hasegawa 1994, 150).

The Comparative Data

This was not true, however, for one of the most important bodies of data on mammalian infanticide outside the Primate order: that from the long-term study of lion behavior in the Serengeti headed by Craig Packer and Anne Pusey (Pusey and Packer 1994). Here again infanticide was emerging as a key explanatory factor in the structure of lion society. One of the key questions about lion behavior was why they, unlike all other big cats, live in groups. Packer, Pusey, and their coworkers argued that infanticide formed a major part of the answer. In order to breed, a male must gain access to a pride of female lions, and his chances are enhanced by belonging to a coalition of (closely related) males. Having evicted the previous male coalition, the males would then set about killing unweaned cubs and evicting juveniles. Despite the difficulty of observing infanticide in a nocturnal animal, ten cases had been seen in the Serengeti by 1990 (Pusey and Packer 1994), and one case had been filmed by a camera crew.[9] This was broadcast as part of the film *Queen of the Beasts,* made by Anglia Television. In other cases infanticide could be inferred or strongly suspected from the disappearance of cubs and the appearance of strange males. In any event, Packer and Pusey estimated that male takeovers resulted in the elimination of almost all cubs less than nine months old. Male sociality could be accounted for by the need for allies in pride takeovers, but female sociality, once assumed to be the product of cooperative hunting (Packer, Scheel, and Pusey 1990), could also be partially explained by the threat of infanticide. Females in very large or very small prides suffer disproportionate rates of infanticide: the only "advantage of group living that we have been able to detect is that moderate-size prides suffer lower rates of infanticide" (Packer et al. 1988, 382).[10] Pusey and Packer concluded that males undoubtedly gained from infanticides and that there were clear female counterstrategies for avoiding it (avoiding the roars of strange males, reproductive synchrony, and so on). From their observations, it was clear that infanticide was a central factor in the evolution of lion society.

Although the example of the lions represented one of the strongest elements in the case for understanding infanticide as an example of reproductive competition, it was only one aspect of the range of comparative literature primatologists and behavioral biologists could now draw on, all working from the basis that infanticide consistently represented an adaptive element in social behavior.[11] This does not imply that it *invariably* had adaptive consequences: in a number of species and studies, such as the chimpanzees or the observations of sea lion and seal behavior (LeBoeuf and Campagna 1994), no clear adaptive advantage could be identified. Nor did it mean that infanticide had to occur wherever the circumstances seemed to warrant it or that the reproductive advantage hypothesis was the only adaptive reason for infanticide, and especially not that only males committed infanticide.[12] But overall there was a clear consensus in the literature that infanticide represented an evolved strategy for increasing individual fitness or reproductive success and that it could be triggered by a range of proximate mechanisms and contexts. The controversy, it seemed, was over—or if was not over, those who continued to treat it as an example of social pathology had retired from the fray. This made it all the stranger that the Erice conference in Italy was marred by deep and abiding concerns about the political consequences of studying infanticide.

The conference, held in 1990, was multidisciplinary, bringing psychological, anthropological, zoological, biological, historical, and sociological data to bear on the effort to understand the range of situations in which infanticide could be seen and its consequences for fitness. Problems arose in the first instance with the inclusion of human data. When it was time to produce the edited proceedings of the conference, one participant withdrew his chapter on parent-infant interactions in a non-Western society out of fear that "the information might be used to further political and economic agendas, such as relocating a tribal group to allow for economic exploitation." Additionally, there was concern that "even though infanticide is a naturally occurring behaviour which is widespread in the animal world, there remain individuals within the scientific community who are opposed to the study of infanticide, and thus seek to block publication of articles concerning infanticide" (Parmigiani and Vom Saal 1994, xv).

Finally, the editors expressed their fear that "political intrusion into the scientific community in general" was becoming more common (1994, xv), although whether this refers to a specific example (such as the growth of the animal rights movement) or a more general concern regarding the influence of postmodernism in the humanities and social sciences[13] remained unclear. This alarm that political elements might either use scientific data for their own purposes or try to restrict academic and scientific freedom to study particular subjects, or to destabilize the scientific establishment itself, led conference participants to produce a statement calling for the protection of the spirit of free inquiry and the prevention of political censorship of scientific results. This statement asserted that just because scientists had found infanticide to be widespread did not mean they condoned it: moral judgments are not to be found in scientific reports. And they emphasized their commitment to this position and the importance of the statement by including their signatures in facsimile in the edited collection (Parmigiani and Vom Saal 1994, xvii).

This almost unprecedented move demonstrates how threatened the researchers felt. But at this stage, as the statement pointed out, infanticide was known to be widespread throughout human history and culture as well as occurring in many primates, mammals, birds, and other species both vertebrate and invertebrate. Why then were they afraid that at this point "scientists will be inhibited from seeking answers or publishing findings" (Parmigiani and Vom Saal 1994, xvii)? Part of the reason might be that the opponents of the sexual selection model of infanticide had found a new champion in Robert Sussman of Washington University, St. Louis.[14]

The *American Anthropologist*

Bob Sussman became involved in the infanticide case for two main reasons. First, his initial interest in academe had been spurred by a passionate opposition to racism, which had left him with an abiding wariness about how scientific research and literature might be abused for political purposes. Second, he was interested in scientific methodology, in the processes for testing hypotheses in the field. Sussman argued that

> if you look at primatology, there's a lot of things that are out there that need to be questioned and need to be tested. Hypotheses need to be tested. Well, this was one of them. . . . In Sarah Hrdy's book, *Langurs of Abu,* there were four cases of infanticide upon which the whole book was based. So, at first I thought, well, it would really be nice to have a review of all the literature, to examine the actual database for this theory. In fact, it was one of the questions I gave graduate students in a class on primate behavior and ecology. And one of the students, Thad Bartlett, actually took me up on it. He did a review of the literature on the infanticide question, and he looked at every case of infanticide. And he asked the question, how many of the cases actually were observed? How many were just plain speculation? And of the observed cases, how many fit the criteria [for the sexual selection hypothesis]? I think he found forty-eight cases in all the literature as reported in our first paper. And of those cases, I think there were eight that fit the sexual selection hypothesis. And of those eight, I believe that two of them involved the father of the offspring. So in fact there were only six cases in the literature . . . that fully supported the sexual selection theory of infanticide.

With his student and another colleague, Jim Cheverud, a population geneticist from Washington University Medical School, Sussman developed this paper into an article that was published in the *American Anthropologist,* the flagship journal of the American Anthropological Association.

This paper argued that a through examination of the infanticide literature was necessary for several reasons. Not only had the sexual selection model become "entrenched" as the explanation for infanticide, but the risk of infanticide was now being treated as a fundamental factor in primate evolution. In addition, though this is implied rather than stated, the large body of data that had built up on infanticide since the mid-1970s had not been subjected to critical review since Boggess's 1984 article. At the outset, the authors acknowledged the difficulty they faced because the published articles now provided so little detail on the *context* of infanticides: wherever possible, they examined the original data, but in some cases they were thrown back on Boggess's earlier examination of the material.[15] Essentially, the article identified two main problems with using the sexual selection model to explain infanticide in primates: the use of the empirical database and the use of sociobiological theory.

As the earlier critics of the sexual selection model had done, Bartlett, Sussman, and Cheverud (1993) provided a chronological review of the database, focusing on cases where infanticide had been *seen* rather than inferred. In discussing the three first "infanticidal" sites, they counterposed the observers' versions of events to Curtin and Dolhinow's—and eventually Boggess's—reinterpretations, showing that there could be alternative explanations. For the later data, where the narrative context was frequently absent, they focused on three main themes. First, again and again they emphasized this lack of detailed narrative and the unanswered questions: Which male killed the infant? Who fathered it? Did the mother become pregnant? Second, they stressed the extreme variability of the circumstances in which infanticide is seen: Were the males new to the group? Were other animals attacked and wounded? How many males were in the troop? How soon after a male invasion did infanticide occur? Third, they repeatedly made the point that the *only* thing these events have in common other than the death or injury of infants is an unusually high level of aggression. However, Bartlett and his coworkers also returned rather more subtly to another major theme of the earlier critics: that infanticide in primates may be restricted to a minority of sites. Using a bar graph and a pie chart, they demonstrated visually how langur infanticides dominate the published records of witnessed infanticidal events: almost half the database is drawn from this one species. Even more telling was that *over* half these reports were from one site—Jodhpur. In fact, Jodhpur accounted for more than a quarter of all authenticated reports of primate infanticide (Bartlett, Sussman, and Cheverud 1993). Implicitly, the audience was invited to conclude that the infanticide literature and thus the support for the sexual selection hypothesis were largely produced by researchers at one site, and one site alone.

This, they suggested, undermined one of the key assumptions made by supporters of the sexual selection model. Just as earlier critics had suggested, rather than being a regular and widespread event occurring throughout the Primate order, infanticide was in fact largely confined to one place. Other assumptions made by the model were also destabilized rather than supported by their examination of the database. In the first case, the argument that infanticide contributed to male reproductive success could not be maintained, they contended, where the identity of the male was unclear and paternity had not been resolved. Simply showing that the female interbirth interval had been decreased demonstrated only that the male *might* have benefited from the act, and in any case, showing that female interbirth intervals decrease after an infanticide was meaningless unless it was compared with the overall infant mortality rate rather than with the interval when the infant survived.[16] The lack of information on male paternity also meant that the selective pressure infanticide might exert on a population could not be assessed. It would be possible to measure this, they argued, if—and only if—the character being measured could be clearly defined. For example, one might decide that if a male is seen to kill infants, he possesses the infanticidal trait: if he is not seen to do so, then for measurement purposes he does not possess it, since for natural selection to work on a trait, it must be expressed in behavior. The fitness rates of infanticidal males could then be compared with rates for the rest of the population. This would be difficult, since male lifetime reproductive success is notoriously hard to study, but Bartlett, Sussman, and Cheverud (1993) argued that this above all was the fundamental test of the sexual selection model. In the absence of this information, all the certainty surrounding the model, and the hypotheses concerning the contribution infanticide made to the evolution of primate sociality, represented, in the words of Glenn Hausfater and Sarah Hrdy nearly a decade earlier, "the construction of sand turrets on sand castles" (Hausfater 1984).

Instead, Bartlett and his colleagues proposed seeking the explanation for infanticide in its immediate, or proximate, context. They argued that the extreme variability they believed was characteristic of infanticidal events, in contrast to the clear patterns identified by supporters of the sexual selection model, resulted from the high levels of aggression produced where competition between males for females is immediate and direct. Rather than targeting infants for attack, the males might just as easily be attacking the mothers, since infants are, after all, usually clinging to their mothers. There were records of males' continuing to attack females after an infant had disappeared, and of males' attacking females without dependent infants. Rather than being a strategy for increasing reproductive success, infanticide might be most appropriately interpreted as an accidental by-product of a situation in which males are competing fiercely for female attention and vulnerable infants are liable to get in the way. In any case, they reiterated, infanticide was extremely rare—only forty-eight witnessed cases existed, and of these only six "fit" the predictions of the sexual selection model. In other cases, such as predation, where a rare event was thought to have had a major impact on primate evolution, there were clear behavioral mechanisms, such as antipredator alarm calls, signifying that predation was a threat. Nothing like this existed for infanticide, they argued. In conclusion, they asserted that in the absence of data on male paternity "there is no evidence to suggest that [infanticide] is anything but a rare and evolutionarily trivial phenomenon" (Bartlett, Sussman, and Cheverud 1993, 985). The similarities between this critique and that submitted earlier by Dolhinow and her students were striking, down to the exclusion of the comparative data—though, unlike them, Sussman and his colleagues did deal with all the reports of *primate* infanticide rather than confining themselves to the langur literature.

No immediate public response was made by those who favored the sexual selection model, not least because of where the article was published. Sarah Hrdy suggested that "the people who really knew the most about [infanticide] were zoologists; they probably didn't even know of the existence of this, and the people who knew the least about the topic were the people reading [it]": anthropologists. As the introduction noted, many primatologists working on the topic were startled to see that issues they thought had been resolved were being resurrected. However, in 1995 Sussman, Cheverud, and Bartlett produced for *Evolutionary Anthropology* a summary of their earlier paper, stating their case in rather more provocative terms. They described the sexual selection model as "a widespread, almost mythological belief, even in the popular literature" (1995, 149) and stated outright their conviction that the wrong sort of data were being used to support the hypothesis. From their perspective, data were being fitted to theory, without a consideration of what data were actually relevant for testing the theory—that is, information concerning paternity and relative male rates of reproductive success. Theorizing and hypothesis forming, they argued, were enjoyable, but "this in itself [was] not science. Good science begins when one collects the relevant data needed to test these hypotheses" (150). The implication was clear: those who argued and wrote in support of the sexual selection model were not doing good science.

Again, supporters of the model were not enthusiastic about responding. Sarah Hrdy recalled that she "wasn't going to reply at all, but Carel [van Schaik] and John Fleagle[17] said we really had to." Carel van Schaik remembered that Hrdy "wanted me to do it instead, but I refused." Eventually, with the help of Charlie Janson, their rebuttal of the arguments of Sussman, Cheverud, and Bartlett (1995) was produced and published as an immediate follow-up to the critique. The tone of their article was placatory. They stressed that they agreed not only that more data from the field and the laboratory would indeed be useful, but that researchers might well have embraced the sexual selection model with rather indiscriminate enthusiasm. But they utterly rejected the charge that there was insufficient evidence to conclude that infanticide was nonadaptive. They considered irrelevant the claim that infanticide was restricted to a few sites: naturally most observations had been made at Jodhpur, since conditions there matched those under which sexually selected infanticide should occur, and langurs had been studied there for longer and in better circumstances than had most Old World monkeys.[18] Equally, the low numbers of recorded infanticides were irrelevant: picking up on Sussman, Cheverud, and Bartlett's reference to the literature on predation, they pointed out that not only do recorded infanticides greatly exceed the number of recorded predation attempts, but behavioral and social mechanisms for avoiding infanticide *do* in fact exist.[19] As for the genetic basis of the infanticidal trait, this had been demonstrated in the rodent literature, which Sussman and his colleagues had chosen to ignore. The example of the lions, where DNA fingerprinting had demonstrated that all cubs born to a pride are fathered by the resident males, gave some indication, they suggested, of the level of selection pressure imposed by infanticide risk.[20] Infants, they emphasize, are regularly targeted when males encounter strange females, and infanticide is consistently found where unfamiliar males replace resident males in a group.

The continuing controversy they put down to one main source: different attitudes toward the proper relation between theory and empirical data:

> While some are interested in emphasising the uniqueness of each case—a valid position—others are driven by the need to seek for general patterns and to use theory to explain them. For the former, it is an insult to the sanctity of the individual and the sacredness of context that generalisations should extend beyond the specifics of the case in hand. The latter derive their greatest pleasure from noting that so many findings could have been correctly predicted on the basis of pitifully incomplete data sets merely by relying on logic, comparisons and extrapolations guided by evolutionary theory. (Hrdy, Janson, and Van Schaik 1995, 154)

Ironically, as in her earlier (1984) prediction that "future generations" would take for granted the adaptive value of infanticide, the phrase "pitifully incomplete data sets" was to come back to haunt Sarah Hrdy.[21]

Widening the Focus

In any case, despite the sidelining of the comparative data, by this stage it was clear that Sussman was determined the controversy would not be confined to the primatological community. Alongside the *Evolutionary Anthropology* article, he had issued a press release asserting that not only was the claim that infant killing produced reproductive advantage greatly exaggerated, but the effect was largely mythical.[22] This resulted in headlines such as the one in the *New Scientist,* the British popular science journal, which told its readers that the "monkey 'murderers' may be falsely accused" (Mestel 1995, 17). Still, it seemed that primatologists and anthropologists were the main interested parties: there was no suggestion that researchers in any other areas of behavioral biology or zoology cared to question the adaptive status of infanticide. But this changed when Sussman became the editor of *American Anthropologist.*

Sussman had been criticized for ignoring the comparative data. In his own defense, he pointed out that

> when I first criticized this theory . . . I didn't think about the evidence concerning rodents and the lions. . . . And I got calls and questions right away (there was a lot of media interest). And the first question that would always come up was, what about lions? And my reaction . . . to this question was, you know, that the evidence seemed to be really good for primates. But you had to look at the evidence . . . you had to examine it very carefully. After examining it carefully, then you could understand why I had questions about it. Well, I hadn't examined the data carefully on lions and rodents.

However, a colleague told him about a woman who had, many years earlier, written a critical review of the data on lion infanticide and had been unable to get it published. As Sussman described it, Anne Innis Dagg, an academic based in the Independent Studies Programme at the University of Waterloo in Canada had

> done work on large mammals in Africa, and [the infanticide hypothesis] just didn't fit with what she'd read about lions, so she went back and examined the data. So when I was the editor of *American Anthropologist,* I said, I want to find out what's happening now, after fifteen or twenty years since she wrote her early article. So I called her and I said, "Would you be interested in reviewing the data from the time you wrote your article to see if the new data would change your mind?". . . She agreed to review the data, and . . . found that everybody, from the time she wrote her early piece to the present, had referred back to the original sources that she had referred to but had misinterpreted the data, repeated these mistakes, or had amplified others' mistakes and never referred back to the original data. It had all sort of built up, like gossip, from that original data and sources.

Dagg's article was eventually published in the *American Anthropologist* in winter 1998. It was the first time this journal had published a paper on the behavior of a nonprimate animal, but Dagg and Sussman believed it was justified by the importance of demonstrating to the audience of anthropologists and social scientists that the comparative evidence for sexually selected infanticide was far too flimsy to warrant applying it to humans in the way done by Martin Daly and Margo Wilson (1988), Desmond Morris (1983), Robert Wright (1994), and other sociobiologists eager to widen their potential audiences.[23]

Dagg pointed out that the infanticide hypothesis "resonates with Western culture, in which many people accept male dominance and aggression and condone in part the control of female sexuality by men" (Dagg 1998, 940). This resonance, she suggested, made it easier for people to accept intuitively a theory that was not well bolstered by fact: close examination of the lion data would show that once again cub killings are the result of aggressive behavior by both sexes during periods of social upheaval and are not evolutionarily significant. In this article, Dagg first defined the sexual selection model in relation to lions,[24] then reviewed the history of the Serengeti lion studies and the cub mortality records in relation to the predictions of the model and explanations for cub death provided by other researchers. So, for example, she approvingly cited the work of George Schaller,[25] who found that the high rates of cub mortality stemmed from maternal neglect. Brian Bertram's suggestion—again in the context of high cub mortality—that lion males, like langurs, might gain a reproductive advantage from infant killing was offset by Dagg's stress on the brevity and relative superficiality of his observation periods along with the paucity of data he could draw on.[26] In contrast, she argued that the work of Jeanette Hanby and David Bygott showed clearly that high rates of cub death were linked with periods of starvation rather than the presence of unfamiliar males, leading once again to the conclusion that cub mortality was caused by maternal neglect or abandonment.

For Dagg the rot set in with the advent of Craig Packer and Anne Pusey, who took over the lion project in 1978 and whose names are most closely linked with the use of the sexual selection hypothesis to explain infanticides in lions. Their accounts of the circumstances and consequences of infanticide in lions are criticized in the strongest terms: Dagg accuses them of excluding information that would have undermined the sexual selection hypothesis and of misinterpreting the data they do put forward to exaggerate how well the sexual selection hypothesis is supported by the facts. As a result, and rather oddly, Dagg largely ignores the work of Packer and Pusey and their research associates in her evaluation of the sexual selection hypothesis, since "the unwarranted assumption that infanticidal behaviour by males is important in lion evolution skews the information provided by later lion workers" (Dagg 1998, 945). Their commitment to the sexual selection hypothesis has apparently rendered their work unusable for her purposes, except where they provide information that might weaken that hypothesis, such as the report that females are also responsible for the deaths of both adult and infant lions. Overall, Dagg weighted the reports of the earlier workers far more heavily than those of later researchers, even to privileging Schaller's description of seeing females mating with male nomads over the use of DNA testing (by Packer and Pusey's team) to demonstrate that all pride cubs were fathered by resident males (Gilbert et al. 1991). Based on Schaller's evidence, Dagg concluded it was not possible to "know with certainty in the field . . . which males fathered which cubs" (Dagg 1998, 946). Earlier researchers, she argued, were unbiased and recognized the role of maternal neglect and environmental degradation in cub deaths: later researchers were blinkered by the sexual selection model.

As in the previous exchange between Sussman, Cheverud, and Bartlett (1995) and Hrdy, Janson, and Van Schaik (1995), the question of what constituted "good science" was raised directly, both by supporters and by opponents of the sexual selection hypothesis. Dagg described the application of the sexual selection hypothesis to lions as "spurious science" where theory is preferred to data in that data are made to fit with theory. She cited Sarah Hrdy's identification of two kinds of researchers doing science in the field: those who focus on collecting data and are "made to sound like unimaginative drones" (Dagg 1998, 947), and those who concentrate on problem-driven research. Dagg suggested there might be a third category, those "who are so enamoured of theory that when the whole of the data do not fit their hypotheses, they ignore basic facts (often produced by the unimaginative drones) in order to salvage at all costs their theory-driven hypothesis. This can lead to fallacious conclusions and is not good science" (Dagg 1998, 947). However, Dagg's own article, her treatment of the empirical observations of lion behavior, and the decision to publish it in a journal devoted to the exploration of human, or at least *primate,* behavior, had shocked and infuriated a significant section of the primatological and zoological community, so much so that they were prepared to condemn it publicly as "bad science."

Joan Silk and Craig Stanford prepared a devastating point-by-point critique of Dagg's paper that was circulated widely before submission to the *American Anthropologist,* accompanied by the signatures of thirty-four more scientists.[27] Kelly Stewart, one of the signatories, remembered her sense of how unusual this response was, since "we like academic freedom; we don't often say, 'this should not have been published, it's irresponsible,' so that shows how rarely an objection like this comes up." The letter, initially addressed to Bob Sussman as editor of the *American Anthropologist,* argued that the paper "is seriously flawed because it rests on invalid reasoning and profoundly misrepresents the empirical literature on infanticide in lions" (Silk and Stanford 1999, 27). Their objections to the publication of Dagg's article were twofold. First, they disapproved of her treatment of the infanticide data, pointing out that the comparative data supporting the sexual selection hypothesis were not confined to lions, as she had implied, and that her analysis of the predictions of the hypothesis in relation to the empirical data from the Serengeti was simplistic, error-laden, and biased. Second, they were deeply concerned about what they saw as the political motivation behind the production and publication of the article.

Again, it is helpful to consider this particular development in its immediate context. Anthropology as an academic discipline, in the United States in particular, had faced serious problems over the 1990s.[28] The unity of the four subfields into which the discipline was divided[29] was under immense pressure as a result of the emergence and development of both "postmodernism" and "sociobiology." In particular, biological anthropologists had been critical of what they saw as a postmodernist turn in the editorial policies of *American Anthropologist* in recent years. In fact, Bob Sussman had been appointed editor at least partly in the hope that he might provide some rapprochement between the biological and the cultural aspects of the discipline (Shea 1999). Certainly, by including an article on lion behavior in one of the very first issues, he was increasing the journal's biological coverage, but not in a way calculated to appeal to all of the biological anthropology community. Silk and Stanford's original letter implied that Dagg's article had been published in the *American Anthropologist* because it would have been rejected by any reputable animal behavior journal.[30] Its appearance, understandably, was likely to widen rather than bridge the cracks in the relationship between the various anthropological subdisciplines. Notably, Sussman and Dagg felt the criticism was ideologically motivated. They both responded with outrage to what they variously called a "political and international mobilization" (Dagg 1999, 20) of hostility and an attempt to "blackball [Dagg] by petition behind her back" (Shea 1999, 25). In her direct response to the (eventual) publication of Silk and Stanford's critique, Dagg expressed a rather disingenuous surprise at the level of critical response her article was considered worthy of, as well as adopting a somewhat careful stance toward the relevance of the comparative literature to the problem.[31]

The topic returned to the *American Anthropologist* in late 2000, when Craig Packer published a detailed rebuttal of Dagg's critique, followed by Dagg's rebuttal of his rebuttal. Packer carefully reviewed Schaller's work (which Dagg relied on so heavily), trying to show that Schaller's observations were *incomplete* rather than *incompatible* with the sexual selection hypothesis. He insisted that their research team had considered alternative sources of cub mortality and had found them negligible. And he drew his audience's attention to the clip of lion infanticide taken from the film *The Queen of Beasts* that could at that point be seen on the Web page of the Serengeti project and invited them to decide for themselves whether this was random aggression or targeted killing. As the study of female counter-strategies and male paternity demonstrates, infanticide as a "fundamental [fact] of social life" (Packer 2000, 830) not only was crucial to understanding lion social life, even where it was rarely observed, but was also crucial to managing lion survival in the context of species endangerment and conservation. He pointed out that Dagg had never studied lions, nor had she spoken directly to those who had, relying on outdated and partial observations rather than considering the Serengeti study in its entirety.

Dagg responded by briefly reviewing the history of the sexual selection hypothesis, concentrating on its lack of empirical support and its inapplicability to human behavior, and attempted to relate the results of the later Serengeti research to the particular methodologies used. Whereas Schaller had studied two prides in depth, Packer and Pusey concentrated on recording physiological data for as many females in as many prides as possible: where cubs disappeared, Schaller, "who knew the mother personally," blamed starvation or maternal neglect, whereas Packer and Pusey blamed incoming infanticidal males (Dagg 2000, 832). Dagg agreed that cub mortality rises when new males enter, and she provided half a dozen reasons (mostly nonevolutionary) why this might be so.[32] More generally, she insisted that the lion data were now unsupported by the primate literature, since that had been thoroughly deconstructed by Sussman and his colleagues, thus suggesting that, if it does occur, infanticidal behavior in lions is atypical. As for the famous film clip of lion infanticide, she argued it was as likely to be the product of the film crew's harassment as the result of sexual selection.[33] Finally, she asserted her right to comment on the lion data regardless of whether she had herself studied lions. Her perspective was broadly based on two key tenets: that it was vital for observers to publish raw data rather than the interpretation of such data, so outsiders could consider the act in its full context; second, that outsiders can often assess the situation more objectively than can those immersed in the problem. Hypotheses about behavior, she concluded, should be accepted only reluctantly and after much testing, since "if they are accepted too readily, they may influence the collection of data, which prohibits the possibility of alternate hypotheses being tested later on" (Dagg 2000, 834).

The Primatological Community

For most of the primatological community, however, the sexual selection hypothesis not only had been accepted but was now a commonplace and fundamental element in accounting for primate behavior. Two wider shifts in the understanding of primate behavior had also been largely accomplished by the end of the 1990s and may well have promoted this acceptance. First, the significance of the subtleties of specifically female social strategies was now widely recognized. Over the three decades during which primate behavior had been systematically studied, there had been a shift in emphasis. Researchers' attention had moved away from the dramatic interactions between primate males within and between groups and toward the more nuanced relationships between females.[34] In this context, an abiding concern with the female strategies that might be expected to have developed in response to the risk of infanticide by males was not surprising; similarly, female infanticide, rare as this was for primates, became less shocking.[35] Second, the tension between socioecological accounts and sociobiological accounts of behavior was also being eased, in a sense, by the developing awareness that while socioecology was fundamental to understanding primate social variation, other primates themselves formed a key part of a primate's ecological landscape (Janson 2000). The theory of primate social behavior was being rewritten to take account of competition between individuals expressed through intra- and intergroup competition and cooperation. Again, the theory of sexually selected infanticide seemed to resonate harmoniously with this perspective.

Leaving the theory aside for a moment, it also seemed that reports of primate infanticide were being made in contexts that buttressed the sexual selection model more and more securely. Infanticide was observed in blue monkeys (Fairgrieve 1995), black-and-white colobus monkeys (Onderdonk 2000), white-faced capuchins (Manson, Gros-Louis, and Perry 2004), and gelada baboons (Mori, Belay, and Iwamoto 2003), to give only a few examples and without including the reports still being made from sites where the act had previously been seen. Increasingly, infanticide was being reported in populations that were known to breed seasonally: that is, where one would not necessarily expect any reproductive benefit from infant killing, since the mother's return to estrus depends on the time of year (Jolly et al. 2000). Perhaps the most significant of these reports was from Ramnagar, in Nepal, a study site developed by researchers from the University of Göttingen. Christian Vogel, Paul Winkler, and others had decided to establish another site for studying langur behavior so they could directly compare their results from Jodhpur. In particular, they were looking for a site where most groups contained more than one male. Carola Borries, one of the key figures in this site's development, was utterly certain that infanticide would not be seen under such conditions: as she said, "I was convinced from the bottom of my heart that in multimale groups you would not have infanticide." However, in the first five years of this study, twenty-five attacks on eleven infants were seen, and one killing was observed.[36] In 1997 Borries published an article showing that males that had immigrated into the group after infant births tended to attack these almost certainly unrelated infants, increasing the chances that their mothers would bear new infants the next year, after these males had had opportunities to mate with the females and therefore father their offspring. Even more dramatically, Borries et al. (1999a, 1999b) were able to demonstrate that males could determine appropriate targets: DNA testing demonstrated that males attacked only unrelated infants. This was the first indication that males really were able to assess paternity in some unknown but effective manner. At the same time, researchers still reiterated that sexual selection was only one of the adaptive explanation for infanticide, which itself was a condition-dependent behavior (Moore 1999).

Increasingly, some researchers treated the notion of infanticide risk as fundamental to understanding primate behavior. The significance of aggression between the sexes, and in a more general sense of harassment by members of the same social group, had been accepted for some time (Smuts and Smuts 1993; Clutton-Brock and Parker 1995; Smuts et al. 1987; Silk 2002), but this was taken one step further by including infanticide risk. Not only was this crucial to understanding female social relationships and the physical distribution of females within a given environment (Sterck 1997, 1998), but it was, as Van Schaik and Dunbar had suggested in 1990, a key factor in the evolution of male-female relationships and the relatively unusual (for mammals) situation in which male and female primates lived in permanent bisexual groups (Treves 1998; Mitani et al. 1997; Van Schaik 2000; Harcourt and Greenberg 2001; Palombit, Cheney, and Seyfarth 1997; Kappeler 2002; Van Schaik and Janson 2000). In a special issue of the *Pro-*

ceedings of the British Academy that reported on a joint meeting between the British Academy and the Royal Society of London to consider the nature of the evolution of social behavior in human and nonhuman primates, Carel van Schaik argued that infanticide must now be considered just as important to understanding primate behavior as ecological factors had always been. Not only did it explain permanent associations between the sexes, but it also explained the vexed question of male social bonding:[37] where the distribution of resources requires females to forage alone rather than as part of a group, a male's interests may be best served by defending a range that covers the territory of many females, and to do so effectively requires alliances with other males (Van Schaik 1996). Van Schaik suggested that this approach might fruitfully be applied to other mammalian species[38]—not least humans. Along with Peter Kappeler, Van Schaik argued that "infanticide prevention ultimately contributes to some of the most outstanding social and cognitive features of anthropoids" (1997, 1692): that is, the evolution of intricate social relationships, whose negotiation requires large, cognitively complex brains. Twenty years earlier, Sarah Hrdy had argued that the risk of infanticide had led to the evolution of what were considered unique characteristics of human females—concealed ovulation and continuous sexual receptivity. At the turn of the millennium, Van Schaik and his supporters had by implication extended the influence of the sexual selection model to what must be among the most fundamental human oddities of all: our large brains and our intense sociability.

Still, voices were raised in opposition. Brotherton and Rhodes (1996), for example, working on dwarf antelopes (dik-diks), argued that in this species infanticide risk—infanticide had never been observed for these animals—was unrelated to the evolution of monogamy. Most interesting, however, was the return to the debate of one of the original key players in the controversy: Phyllis Dolhinow. Now professor emerita at Berkeley and accompanied by a new generation of graduate students, Dolhinow was proposing a new explanation for infant killing. Just as researchers were now reaping the full benefits of treating female primates as independent and active agents, she proposed that it was necessary to regard infants in the same light. This was not a new position for her: in 1993 she published an article that returned to the old mother-infant separation experiments conducted at the Berkeley langur colony in the 1980s, which emphasized infants' initiative in relation to adult interactions (Dolhinow and Taff 1993). Other researchers (Pierotti 1991; Gomendio and Colmenares 1989) had also regretted that most explanations of infanticide "ignore the possible role of the offspring in the interaction and treat offspring as if they were passive recipients of their fate" (Pierotti 1991, 1141), but this was the first sustained attempt to argue that infants were, to a marked extent, responsible for the fatal events observed at some study sites.[39] The Berkeley colony had provided immensely detailed records on langur physical and social development, and Dolhinow was able to use these to demonstrate that, overwhelmingly, the infant or juvenile initiated interactions with adults, especially with males, and that these interactions, especially when new males were introduced to the colony, frequently led to aggression between the new males and the colony females. These observations, she argued, demonstrated that "young langurs can be endangered in situations that arise in part from their own and their mothers' actions": frequently, the male is a "passive figure" until forced to take action (Dolhinow 1999a, 192).

Conclusion

Of course infants and juveniles that are old enough to move independently are largely irrelevant to the sexual selection hypothesis, which requires that the male direct his attention only to dependent infants. However, Dolhinow's account of langur behavior treated this interaction as a possible solution to the "langur murder mystery" (1999) —a solution that escapes what she identified as the prevailing and unfortunate bias toward providing adaptive accounts at all costs. In this she was echoing Bob Sussman's position. In his multivolume primate behavior textbook (Sussman 1999; 2003), he also condemned what he considered to be the rush to theory in the absence of substantive data. But the infanticide controversy in its various phases was about far more than a hypothesized preference for theorizing without data or an inability to see beyond the confines of individual investigations, as part 3 will show.

It is clear that by this stage positions within the controversy had become so entrenched that some intimated bad faith was involved. Participants increasingly saw ideology as intruding into scientific debate: the controversy was no longer about how and why one should study primate behavior but about one's wider political position and the willingness to compromise scientific standards in order to perpetuate them. But once again this was not the whole story. For supporters, the fieldwork that had contributed to the stabilizing consensus since 1974, around the sexual selection hypothesis in particular and the adaptive account of infanticide in general, represented almost unassailable grounds for accepting the hypothesis. Equally, for those not so disposed, neither the primate data, nor the comparative data, nor the experimental work made it possible to consider the hypothesis either proved or not proved: for them it remained ultimately untested. One could argue that it was exactly this situation that the earlier and intense methodological discussions among primatologists, deeply aware of their multidisciplinary origins and the relevance of their subject to the question of human identity, had hoped to circumvent. For the most part they had succeeded and avoided costly and interminable debates on the interpretation of behavior. Then what was it then that made the topic of infanticide so recalcitrant?

PART THREE

Questioning the Field

SIX

Accounting for Infanticide, 2001–3

Introduction

I think initially they had a good point. Hey, most of you who are claiming infanticide are assuming infanticide. And yes indeed, I think people began to be a lot more careful about how they described whether they'd seen infanticide or not. . . . So in that sense the criticism . . . made people advancing a new theory a lot more careful with their evidence. Good, that's exactly what we need. But I would have said right from the start also, people were aware that they were being overcritical. That when there's a female and a strange male in a bush and the female comes out with a dead infant . . . the fact that you didn't see it doesn't really mean the male didn't do it. . . . The demand for evidence I think was good. Their claim for lack of evidence was simply exaggerated. . . . Some of the criticism of the SSI [sexually selected infanticide] hypothesis is that you have to know fully what's going on and you've got all the data before you can produce a hypothesis for it. What kind of hypotheses do we get excited about? Those that explain something new, that need more data to show whether they're correct or not. It's a misunderstanding of how scientific bright ideas occur.

Alexander Harcourt, *University of California–Davis*

Hypotheses need to be tested. . . . In the end, their data is based on very slim field data or captive data, in which they look at the specific cases. They forget about the whole context of gene-environment interaction and demography. How many young die altogether, and how do the young that die that aren't infanticide victims affect the interbirth interval, and statistically, does that make any difference? There's never any attempt to actually answer that question.

Bob Sussman, *Washington University, St. Louis*

Made nearly thirty years after the controversy surrounding infanticide in primates began, in many ways these two statements encapsulate the issues at stake in the debate as it stood in the early years of the twenty-first century. Both speakers are scientists, both have vast experience of primatological fieldwork, both are committed to working within the Darwinian framework of evolutionary biology. On one reading, the two are in agreement: science proceeds by testing hypotheses against the appropriate evidence. But on another, it is clear that there is fundamental disagreement about what might constitute "appropriate" evidence and how such evidence might be adequately tested. From both perspectives, hypothesis construction represents a driving force behind scientific innovation and the advancement of knowledge: hypotheses are tested against an expanding body of information that will in turn inspire the development of new ideas. However, while one point of view focuses on the immediate context of the infanticidal event for the individuals involved, the other insists on interpreting that event against a far wider socioecological setting. These perspectives are by no means incommensurable, and in all probability both researchers would attest that both are central to understanding primate behavior, particularly behavior in the field. But it is in the particular context of field science that apparently inconsequential differences in the stress laid on particular aspects and areas of scientific practice can be magnified into full-blown controversies such as that relating to infanticide—as is evident from the way researchers from both sides of the debate discussed the ways infanticide might be studied in the field.

In this chapter I will examine the accounts primatologists gave of how and why the infanticide controversy developed as it had over the past three decades. Among other things, I asked researchers to explain why the controversy had emerged in the first place, why it had evoked such passions, and why resolution had proved so impossible. I asked what evidence they would need before concluding that an act of sexually selected infanticide had taken place and how one might go about investigating the origins and consequences of such a rare act under both field and laboratory conditions. The answers given to these questions were intriguing.[1] Not surprisingly, respondents explained the advent and persistence of controversy by referring to contingent factors ranging from the existence and exacerbation of personal hostility between individuals to the influence that wider intellectual movements, such as the rise of both sociobiology and postmodernism, had had over the natural and the social sciences. Less predictable, however, was the suggestion that the "science wars" had themselves had a role in the controversy: particularly for a historian of science, the suggestion that the (sometimes heated) debates between students of science and scientists in the middle to late 1990s had actually affected the course of a scientific dispute was somewhat startling.[2] But most fascinating was that descriptions of the way infanticide should be defined and studied in the field revealed profoundly different interpretations of the problems of fieldwork and of how one should go about doing science in the field when studying the behavior and interactions of long-lived and highly social wild animals.

I will begin to explore this issue by examining the way scientists argued that the infanticide hypothesis should be defined and investigated in the field. I will also consider the contingent issues that participants discussed in explaining why infanticide had become controversial in the first place and why controversy had persisted, For the most part, however, this chapter will focus on the way scientists described and explained how one should conduct fieldwork and investigate rare behaviors in the field (and how they themselves went about it). In this context they discussed the importance of *prediction,* the role of *observation,* and the relation between *theory* and *data* in studying evolution in the field. But to begin with I will consider the role participants assigned to the most obvious and familiar means for testing hypotheses—the *experiment.*

Testing Times

Ethics, pragmatics, and primatological tradition all forbade experimenting with infanticide in the field. As chapter 1 described, primatologists have long been wary of manipulating the behavior of the animals they watch, and in the case of infanticide, none of the researchers I talked with were willing to countenance any kind of invasive experimental manipulations that could possibly result in the death of free-living animals. For Kelly Stewart, based at the University of California–Davis, even in species that local human populations treated as vermin, "it's just considered unethical. I mean yes, baboons are shot as crop raiders and pests, but to do it simply out of . . . scientific curiosity just isn't good enough." While playback experiments were barely acceptable,[3] removing or introducing males in the wild was not. As Charles Janson, from the State University of New York at Stony Brook, said, "We're happy to fiddle with vocalizations, and we're happy to fiddle with predators, sort of in the abstract, and we're happy to fiddle with food availability, because none of those things induce permanent changes in the social system." Louise Barrett, now at the University of Lethbridge in Alberta, took this one step further, suggesting that if one were to intervene in the life of a "highly social group living animal, you're going to interfere with lots of other things besides just your experimental conditions. . . . You might then be distorting a whole bunch of other variables that you can't know about if you're going to do it in a field situation." If one were to interfere in the social system itself, far more questions would be raised than answered: unlike the laboratory where, at least in theory, all conditions other than those being directly manipulated can be held constant, the field is a far more fluid and potentially chaotic environment. Primatologists had come to the field in the first place to study the ordinary behavior and relationships of the individuals belonging to primatological groups: intervening in these systems risked changing or even destroying the social systems they had come to study.

Fundamental elements of field practice when studying social behavior also militated against manipulation. Most primatologists, especially where behavior was being studied intensively, did not examine whole populations of primates but focused on individual groups within that population. Habituating animals to human observation is both laborious and time consuming, and realistically any one person can follow only a single troop or animal at any given time. Therefore most primatologists study a very small number of troops, each representing a single data point. Even if one attempted an experimental manipulation of these animals' behavior, it would still be extremely difficult to make statistically valid generalizations from the results of such interventions.[4] Andreas Koenig (State University of New York at Stony Brook) pointed out that "in order to have a valid sample, you have to do this with ten groups, and . . . most people don't study ten groups, they study two or three groups. . . . So, you never get [an adequate] sample size. If you want to do this, you have to do this with a larger sample size, which is not feasible, and then comes this conservation aspect on top of that."

Universally, then, researchers found the idea of conducting experiments that might in any way endanger the lives of primates not only morally repugnant, but also simply impractical. If one were to interfere, the results of the manipulation would be open to challenge not only on the grounds that it was impossible to draw definite conclusions about the proper relation between cause and effect in a complex and imperfectly understood social system, but also on the statistical validity of the sample size. By doing so you might well also have ruined the prospects of further work with a group that it had taken effort and time to locate and habituate. It seems clear that free-living primates are not regarded as good subjects for experimental manipulation.[5]

Rather than turning to experiment, the supporters of the hypothesis generally agreed that the way to approach infanticide in the wild was to examine the behaviors surrounding it. But this meant drawing on a range of evidence representing different degrees of certainty and reliability. For example, one might consider the immediate behavioral context of a potentially infanticidal episode: Sarah Hrdy pointed out that "when males go into an infanticidal mode they display a very distinctive and intense state of arousal with an erect penis, vocalizing with a cackle bark which you hear in almost no other context, and of course, all the stalking." To supporters of the adaptive model, all these behaviors characterized an infanticidal event.[6] Another source of evidence could be found in the wider patterns seen in the primate group: the relationships between males and females, juveniles and adults in different structures of primate society.[7] So, for example, Kelly Stewart argued that you could "ask questions around the infanticide phenomenon. For instance, how much has it shaped relationships within the group? How much has it shaped male-female relationships? And so you formulate a hypothesis, you say, well, if male protection is really important, then females should be more responsible for [maintaining the relationship], and that sort of thing." Perhaps more directly, now that noninvasive DNA testing could be effectively carried out in the field, it was possible to compare the way males behaved toward related and unrelated infants: Louise Barrett cited a recent paper demonstrating that males directed more care toward their genetic offspring than toward other infants they associated with and asked, "If you can do it that way round, if you can detect who is yours, somehow, then surely it's also possible that you can detect who isn't yours?" All these strategies, however, were focused on the immediate context of infanticidal events, rather than on tests of the hypothesis itself, and hence their validity relied on how the observer was willing to interpret the data: a key question, as later sections of this chapter will show. For example, would an observer hostile to the sexual selection hypothesis have recognized Hrdy's account of "characteristic" male infanticidal behavior?

Direct tests of the sexual selection hypothesis were also available, however. For example, one could take a cross-site comparative and statistical approach, as chapter 1 suggested, combining the data for all known or suspected infanticides and using them to test competing hypothesis, as Dunbar and Van Schaik did in relation to the origins of monogamy in primates, and as Paul Newton did with the infanticide debate (Van Schaik and Dunbar 1990; Newton 1988). This was a relatively arduous task, and clearly also subject to interpretation, since it would rest on a rereading and reclassification of work carried out by other people at sites distant from one's own. In contrast, for some researchers the only thing needed to show that infanticide was occurring as a result of sexual selection was the demonstration that the interbirth interval for the mother had been reduced and that the lifetime reproductive success of the male had thus been increased. In this sense it would not even be necessary to show that the *comparative* lifetime success of the male had increased (that is, that he left more infants than a noninfanticidal male might). As Carola Borries (SUNY, Stony Brook) put it, "If that male has an advantage, regardless of his lifetime reproductive output, even if that's still small, or smaller than average, it is still more than he would otherwise achieve." In this sense, as later sections of this chapter will show, complete and detailed observations of infanticide, whether naturally occurring or experimentally elicited, had become only one aspect of the investigation of the consequences of a complex—and condition-dependent—behavioral trait.

Notably, the unwillingness to test the sexual selection hypothesis directly in the field did not mean experimental data for the sexual selection hypothesis were completely absent. Researchers working on captive rodents had been able to evoke and study infanticide under controlled laboratory conditions. These data had been eagerly followed by primatologists such as Sarah Hrdy, who remembered that "as soon as I found out about the rodent work, I started to work very closely with those people, and to be very influenced by them, and to appreciate that work a lot. Because I thought it was going to solve that problem." Similarly, Tim Clutton-Brock (Cambridge University) and Robin Dunbar (Oxford University) both explicitly made the point that infanticide had been thoroughly examined under controlled laboratory conditions, with the result that "there is detailed experimental evidence in rodents that it occurs under more or less the circumstances that you would expect under the sexual selection hypothesis" (Clutton-Brock). Problems remained, in that laboratory conditions are by definition unnatural and thus left open the question whether sexually selected infanticide would occur in a less artificial situation, but there was still a strong sense that the rodent data in particular could effectively complement the observations made in the wild. That is to say, the sometimes fragmentary observations of infanticide in the wild could be balanced by complete, timed, and constantly monitored descriptions of events in the laboratory, while the artificial conditions that prevailed in the lab would be compensated for by similar behaviors in unmanipulated populations of free-living primates. But as some commentators noted, the validity of generalizing from rodents to the primate situation remained unproved.

The obvious solution might have been to study infanticide in captive primates—but this prospect also raised immediate ethical and practical problems. Even with the rodents, the researchers had had to stop the male from actually killing the infant;[8] in primates this necessity was even more pressing. As Leslie Digby (Duke University, Durham, North Carolina) put it, "You can't study the phenomenon itself, only the things leading up to [it]: increased aggression, increased interest in the other infants." Alternatively, you could, as Carel van Schaik, now at the University of Zurich, suggested (with the caveat that he himself would be reluctant to do such a thing), try something along the lines of the rodent experiments. That is to say, one could "introduce a new male, take out the old male, introduce a new male and somehow there, there's some mesh, a barrier between them, and then you study the behavioral responses." But he and others emphasized that such studies would be very tricky to do. Not only would they cause great stress and tension to the animals, but it would be very difficult to decide how to interpret events if infants were not killed. As Van Schaik explained, because "you don't want to endanger the lives of the infants . . . every experiment you do will have its limitations, in that you cannot observe the actual attempt at infanticide and the responses to that. Because you know, it's a little bit like doing those predator experiments . . . by presenting model predators. Well, the first steps are perfect. But then, of course, the predator doesn't do anything, and so everything that follows is questionable." In the absence of infant death, one cannot be certain that infant death would have followed, but ethical considerations preclude an experimental protocol that would endanger them.[9]

Not surprisingly, opponents of the sexual selection hypothesis challenged the relevance of the experimental rodent data to the primate situation, and not only because of the artificial conditions in which the animals were kept. For Bob Sussman, since "most of the stuff . . . on rodents is from captive populations, which are highly genetically bred," there is a serious question whether one can make any generalizations from the social behavior of these artificially produced animals. Jim Cheverud, who as a population biologist at Washington University, St. Louis, had extensive experience with laboratory animals, felt that the rodent work "wouldn't speak to the primate problem for me, especially when it comes to behavior," because of the assumptions one must make when generalizing from one species to another. Even at the physiological level, when rodents are studied as direct substitutes for human bodies, "there will be some things that we will find are very important in rodents that have no role in humans at all." At the far more complex behavioral level, correspondences are even more conditional—and as Cheverud argued, "in labs, mice will do terrible things, that I don't imagine happen a lot in the field." Though the experimental study of primates was unacceptable to him, so too were the rodent data. The genetic manipulation of laboratory animals, the dangers of drawing too hasty comparisons between the behavior of individuals belonging not just to different species but to different families, and the likelihood that captive animals will exhibit pathological behaviors all meant that the opposition rejected the rodent data as an experimental demonstration of the predictive power of the sexual selection hypothesis. In fact, for the opposition, the comparative data were in general far less relevant to the infanticide debate.[10] Again, this was not a novel situation for primatology in general, where evidence from other animal species was rarely cited.[11]

Patterns Of Prediction

As significant as the comparative data were to supporters of the sexual selection hypothesis, however, of equal moment was the sheer predictive power of the model. Every researcher considered infanticide an event that was both infrequent and difficult to observe—the reason supporters of the hypothesis were prepared to accept evidence implying infanticide as well as fully witnessed attacks as buttressing sexual selection. As Louise Barrett put it, "The best kind of evidence is to see the actual act take place," but circumstantial evidence such as "a dead baby with wounds on it that was perfectly fine yesterday" combined with the arrival of a new male would be very strong grounds for suspecting infanticide. Similarly, Phyllis Lee, now at Stirling University in the United Kingdom, argued that "in order to call it an infanticide, you either have to have good visual grounds or good post hoc dead baby grounds, with the wounds on it obviously caused by male canines," and Kelly Stewart felt that "if you find an infant dead on a trail and there's no evidence of [an interunit] interaction, you don't count it as infanticide." In other words, at the most basic level, infanticide involves both contact between a strange male and a group and the death or disappearance of an infant. While witnessed and very strongly suspected cases of infanticide should be kept separate for analytical purposes,[12] these two factors are central: the primary association between the appearance of an unknown male and the disappearance of an infant member of the group. In the absence of either element, infanticide has not taken place, but in the presence of a new male, infanticide is predicted to occur.

But at the same time, researchers were keen to emphasize that not *every* instance of infant disappearance or infant killing had to be the result of sexually selected infanticide—and to stress that mistakes were perfectly possible. Some supporters of the sexual selection model thought controversy might have persisted because there had been "too facile an acceptance of a hypothesis," in that there was "a willingness to see every infant death as being [explained by] the sexual selection hypothesis" (Kelly Stewart). Carel van Schaik emphasized that infanticide "has multiple explanations, as everybody has stressed in the past, and we still do now, sexual selection doesn't explain it all [and] not every individual case has to be adaptive because animals aren't designed to be that way," echoing a point made by Phyllis Lee, who argued that while infanticide was undoubtedly adaptive, it was still "going to occur in some conditions where the male makes mistakes." The question was whether the strategy succeeded overall: as Charles Janson put it, "We make generalizations all the time without expecting them to apply 100 percent to every instance of something we see. . . . I would even be willing to accept the idea that a certain fraction of infant killing that you see out there is accidental . . . but that explanation doesn't end up producing that many interesting predictions, whereas the sexual selection hypothesis does." Again, the predictive fertility of the hypothesis is the source of its appeal. In the sexual selection hypothesis researchers have found a means of making sense of the world, not invariably or universally, since it deals with the behavior of fallible animals, but *most of the time.*[13]

Again and again, supporters of the sexual selection hypothesis stressed the importance of prediction: that infanticide could be *predicted* in given situations gave basic grounds for accepting it as an adaptive rather than a pathological act.[14] As Carel van Schaik said, "It doesn't really matter how many infants get killed, if it's predictable enough, then . . . it becomes the target of selection." It is the predictability of events rather than their rate of occurrence that is essential here. In this sense, the significance of observing an act of infanticide has changed profoundly. That infanticide has been observed under conditions that match those predicted by the sexual selection hypothesis is still fundamental to acceptance of the theory, but what is of ever greater interest to researchers is not so much recording the act itself as the influence infanticide risk has had on the evolution of social systems in general and on relationships between male and female primates in particular. In this sense, for example, whether an act of infanticide has been observed for a particular species can be less important than whether infanticide-avoidance strategies can be identified in patterns of social behavior.

At the same time, supporters of the sexual selection hypothesis were emphatic that personally witnessing infanticide was critical to an individual's assessment or acceptance of the theory. So, for example, several researchers said they had been reluctant to witness this behavior: before traveling to

Ramnagar, Carola Borries remembered, "I hadn't seen it, and I also wasn't very keen on seeing it"; Volker Sommer, now at University College London, stressed that "I'm a very vigorous defender of what I saw, because I didn't want to see it. I didn't go out there to see it." That they were reluctant witnesses adds weight to the significance of their testimony, as far as they are concerned. But it seems that seeing infanticide makes the sexual selection model far more plausible. Carel van Schaik described how, earlier in his career, he had worked with captive primates in Utrecht, "where we had to put in new males once in a while, and before there was enough space to . . . make soft transitions[15] we'd all be standing there with the fire spout, and as soon as the guy started to be nasty, we'd drown him. . . . We really managed to minimize infanticide that way. But you see, it's those experiences . . . that make you convinced that there's something there." Seeing, it appears, really is believing. Supporters of the hypothesis were almost unanimous in affirming both their revulsion at the events and their conviction that the attacks were deliberate and targeted.

But this last point implies another: one cannot observe behavior without at the same time interpreting it. So, for example, Thelma Rowell[16] mirrored Carel van Schaik's argument when she explained that she was opposed to the sexual selection model in part because "I've sat in the field and watched monkeys probably as much as anybody else, and I never saw anything that remotely made me suspect that infanticide was even a gleam in the back of the mind, you know. I'm sure that if I had seen an incident, and I don't doubt that they occur, I might well be affected, but I never did. Even working on species that are supposed to do it."

It isn't simply that she never saw infants being attacked; she never saw any activities that might possibly be associated with such attacks. The question remains open: Would a researcher predisposed to accept the sexual selection hypothesis have interpreted these activities differently?[17] Andreas Koenig made the point that "even if [people] see the same situation, they may perceive it differently," an event he had directly experienced at Ramnagar:

ANDREAS KOENIG: We had this one case, one student . . . coming to the site in Nepal, and he was not seeing a male attacking an infant.

CAROLA BORRIES: But it was happening in front of him and he couldn't see it.

AMANDA REES: What was he seeing?

ANDREAS KOENIG: Nothing.

CAROLA BORRIES: Nothing was happening, they [the langurs] were just running around.

Two people watching the same thing will give different accounts of what they saw: interobserver reliability can be high, but it can never be complete. Where there is fundamental disagreement about the possibility that a (rare) behavior might occur, the way fragmentary events are interpreted becomes central.

This is important, since it directly concerns the nature of observation, the relation between theory and data in a field science, and the predictability of infanticide. Nor is it a new problem for primatology: a key challenge for field-based primatological research since its inception has been understanding variability in primate behavior. At the outset, the traditional ethological presumption that there was such a thing as "species-typical behavior" had meant that differences observed between populations were ascribed to human error of one kind or another. Gradually, the extreme variability of primate behavior under natural conditions was documented and ascribed first to ecological conditions and later to evolutionary pressures. But it was far harder to do this effectively for rare behaviors (infanticide, tool use, meat eating, and so on) than for more mundane activities. After the sociobiological revolution, this problem became both less relevant and more acute, in that the quest to establish species-typical behavior had become less important at the same time that a novel, though sometimes controversial, source of hypothesis construction became available. In this context, sociobiological theory itself provided a functional alternative to experiments in the field: researchers could now go to the field with specific aims and questions in mind, examining their data in a comparative and cross-site context as a means of subjecting their ideas to "natural" experiments. However, individual researchers record—and as an adjunct to that recording, interpret—data in very different ways; and depending on their particular theoretical background, they will define certain aspects of primate life as relevant while excluding areas that other researchers might regard as central. I will discuss this in more detail later, but for now it is sufficient to note that the infanticide data were recorded by many individuals under observational conditions that varied tremendously. For cases to be reported as examples of adaptive infanticide, observers must have evaluated the things they have seen against the requirements of the hypothesis and found them acceptable. Another researcher seeing the same set of events might well not have done so.

Certainly, opponents of the hypothesis have used the database to demonstrate that *very few* of the accounts of infant injury, death, or disappearance actually correspond with the predictions of the hypothesis in its *entirety.* Moreover, as became clear in the course of discussions, for the opponents, fitting the pattern of events to those predicted by the hypothesis is merely the first step in testing the assertion that infanticide has developed among primates as a sexually selected strategy. Bob Sussman believes that those who support the sexual selection hypothesis are missing the point of his comments:

> When they say, well, we've fit these criteria that these guys have set up for us, when in fact they [the supporters of the hypothesis] originally set up the criteria, and we said, "Yes, you really have to [adhere to these]. If your cases don't fit these criteria, they are eliminated immediately," but even if particular cases do fit, then you have to ask, "Does it make a difference evolutionarily?"

For Sussman and his colleagues, demonstrating that a male had killed an infant in circumstances compatible with the sexual selection hypothesis (that is, the male was unrelated to the infant and went on to mate with the mother, who in turn showed a reduced interbirth interval) only confirmed that an interesting question needed to be investigated. If one wanted to actually test the hypothesis, it became necessary to examine a number of other sources of data. First, one had to identify the immediate, or proximate, factors that triggered the event. Second, one had to show the biological (genetic) basis of the trait that selection is acting on. Third, one had to measure and compare male relative fitness over time within a population to show that infanticidal males were leaving more offspring than noninfanticidal males. From the perspective of the critics, *only* if this final qualification is met is it possible to say that infanticide is a sexually selected strategy. As Bob Sussman put it:

> What's the hypothesis that they're stating? Basically, they're stating that there are males that commit infanticide and those that don't. These males that commit infanticide have a genetic trait or a group of genetic traits that enable them, or drive them, or are underlying factors that cause these males to kill these infants, right? It's a genetic factor. If it wasn't genetic, it wouldn't involve selection. So there are males that have an infanticidal gene or genes, who knows, and males that don't. There has to be a differential in fitness between these two. . . . Second, they're saying that these infanticidal males are reproducing more than those that are not infanticidal. And third, they're saying that over time, over many generations, the infanticidal males are leaving more offspring. . . . So another claim is being made, not only that there is a genetic trait for infanticide, that there is differential fitness in these traits,

> that these males are leaving more offspring, but also that evolution is occurring in every generation in these populations. There is an increase in the frequency of infanticidal genes each generation.[18]

From this perspective, those who defend the adaptiveness of infanticide have missed the point. It isn't so much the problems with the data that already exist (the partial or complete observations of infant death or disappearance) that make it difficult for opponents to accept the hypothesis;[19] their concern is, first, with what they consider assumptions the hypothesis makes about the wider historical and demographic context of infant killing, and second, with whether infant killing is currently affecting the genetic makeup of the population.

Aggression in Context

Ultimately for the critics of sexual selection, a robust test of the hypothesis would require one to "examine those males that have committed infant killing . . . and see if they have contributed more offspring to the next generation than the ones that haven't. And we would need to do this over many generations. That's how we measure fitness" (Sussman). Jim Cheverud warned that just measuring male relative fitness was not necessarily sufficient, since "it also makes a difference when in your life you father them." Instead, measurements of male lifetime reproductive success would be necessary, since the important factors in male dominance are "age and how long he's been in the group, not heritable features. Because the young animals he's dominating now and that are not doing the breeding, some of them are going to grow up to be him, while he's off losing his teeth." In a way, Cheverud is suggesting that part of the problem in studying the evolution of primate behavior comes from the way field studies are structured. Over the course of the twentieth century, the ideal field study grew steadily longer (Rees 2001a, 2006a), but finding funding and support for such long-term studies is difficult. As a result, many still take a cross-sectional rather than a historical approach, which means that "not all of the males are equal . . . they're not comparable, they're not [being studied] at the same time of their lives" (Cheverud). Additionally, to test the SSI hypothesis one would need to know more about the wider relatedness of the population under observation and its general patterns of morbidity. Thelma Rowell pointed out that in "real life, if a monkey gets to produce ten babies in her lifetime, she's doing extraordinarily well, and very few of them do that much. Which means that anything which goes against that ability to produce that small number of offspring . . . seriously limits what males might do by way of exploiting females to their own advantage, because they do after all have mothers and sisters and daughters." For her, one of the fundamental problems with the hypothesis was that it assumed "that these males arrived from outer space with an agenda," without considering what the relatedness might be between these strange (to the observer) males and the known group animals.[20] Finally, the numbers of infants killed (or thought to be killed) by males had to be placed in the context of the total rates of infant death for a population. Infant death from whatever cause would reduce the interbirth interval. To quote Bob Sussman again,

> They think they've proven the sexual selection hypothesis if they can show that these males did indeed have offspring after they committed infanticide. What they don't know is, How does that affect the whole population? . . . If there's fifty deaths and five of them are caused by infanticide, how are these other forty-five deaths distributed among the population? . . . How can these five infants have anything to do with what's happening when there's, say, one hundred infants born, and often over 50 percent of young die from various causes before they reach maturity?

This point is of central importance since, as this chapter has shown, it is the reduction in the interbirth interval that supporters believe provides important evidence for the hypothesis.

In the absence of these data, critics of the hypothesis hold to their own explanation for the events at sites where infants were seen, in whole or part, to be attacked or killed by adult males. This explanation focuses on the place of aggression in primate social life and the possibility that infant killing may be an epiphenomenon produced by selection for aggression. Putting it succinctly, Sussman suggested that "under certain circumstances, males get aggressive, and under those circumstances, they're more likely to kill an infant than not." Citing captive research demonstrating that males' testosterone levels rose sharply when they encountered unfamiliar females, or females they had been totally separated from for some time, Sussman suggested that one should "think of these males that are coming in in influxes, that have been separated throughout the whole year. . . . All of a sudden they join groups with females, and their testosterone levels rapidly increase and, along with this, so do their levels of aggression. This is a natural phenomenon. Furthermore, Phyllis Dolhinow observed that langur infants are really attracted to any kind of action and want to run to their mother when anything's happening, and so it makes perfect sense" that infants are killed in these circumstances. Explanations for infanticide, especially in the context of the problems with the database that render the claims for predictability rather less powerful, should be sought in the immediate social and physiological context produced when a new male enters the group.

When one does turn to ultimate factors, it may be that selective pressure has been brought to bear not on infanticide specifically, but on aggression in general. Considering the exploration of ultimate factors, and emphasizing his primary affiliation as a population geneticist, Jim Cheverud pointed out that neither characters nor traits could be examined in isolation, as he proposes that those who support the sexual selection model are attempting to do: "If I wanted to study the size of the face, in evolutionary studies, I wouldn't just study the size of the face. I would look at it in anatomical context. I'd have to study the braincase and other things that I know will interact with it, because otherwise you can get a very false impression of what natural selection is doing."

In other words, what appears to be selective pressure on one trait may well be a result of pressure on another: selective pressure on arm length will produce a correlated increase in the length of the legs, even without direct pressure in that direction. If this can occur at the anatomical level, then it may well be even more likely to occur at the more complex behavioral level. From his perspective, rather than singling out the elements of the interactions that the sexual selection hypothesis treated as significant, infant killing had to be examined in a wider evolutionary context. He conjectured, "I would think that you would get more infant killing when you have a more aggressive species. That would [make infant killing] a correlated response to the general level of aggressiveness that's been selected for in the species for its social interactions, rather than seeing [infanticide] as the prime mover of the evolution of everything in social groups." Infant killing occurs when tensions and aggressions in the group are already high, and infants are more vulnerable to attacks than are adults. Rather than the risk of infanticide raising group tension, it is more plausible from this perspective to consider infant killing the result of elevated levels of aggression.

However, it is rather hard to imagine how infant killing—by definition an act of aggression—could occur in a nonaggressive situation. Louise Barrett argued that

> [the problem] with Bob Sussman's argument is [that] he always talks about, oh, it's just a by-product of male aggression, and I think, well, it's got to somehow involve male aggression, it's an aggressive act. . . . And if what you see is that males come into the troop, and they are aggressive, they're hyper,

> they're full of testosterone, and they go 'round being aggressive to everybody, that's true, but . . . the infant is designed to be cute so that everyone will look after it . . . and the male's . . . got to overcome [this] and what better way than to get utterly revved up and aggressive and then do it? He's saying, well, it happens as a by-product of males just being generally aggressive but . . . the proximate mechanism has to be something to do with getting them completely revved up and aggressive so that you can do this thing. You can't just walk in and go, oh, that's not mine and just casually bite its head off in a nonaggressive manner.

This was echoed by other supporters of the hypothesis, who argued that Sussman was demanding the impossible: How could one have a nonaggressive aggressive act?

His other proposed tests of the SSI hypothesis were treated as equally unworkable, especially in the field environment. Invariably, when asked about Sussman's three tests (identifying the biological basis of the infanticidal trait, establishing that it was differentially distributed throughout the population, and demonstrating that males that possessed it left more offspring than those that did not), those who accepted the SSI hypothesis argued that in relation to this particular problem, either the standards of evidence were being set too high—that a double standard was being applied—or that the question itself was irrelevant. So, for example, Charles Janson agreed that "it would be lovely to have evidence of this kind, but I would argue that if that's the standard of evidence for field biology, then no field biologist has ever proven anything to be adaptive, because we don't have that evidence for any trait I know of, in behavior, in the field." Leslie Digby pointed out that "all of those things would be lovely, but I don't think we've demonstrated that for the shape of the hand, which I would say is pretty adaptive, or the shape of the hip, which is a unique human trait that is adaptive for bipedal walking, yet we haven't traced that to a genetic component. . . . So it's not only holding it to a high standard, it's holding it to a standard that is beyond our ability right now." Essentially, in demanding that supporters of the hypothesis demonstrate the biological base of the infanticide trait, Sussman was refusing the phenotypic gambit,[21] a strategy evolutionary biologists used to allow for "a leap of faith, of necessity, . . . from proximate mechanisms to fitness, because invariably, you can't measure fitness for most organisms unless they're very, very short-lived, partly because it requires a minimum of five generations to measure fitness and if you're dealing with long-lived species, that's simply not practical," as Robin Dunbar put it.[22]

Emphatically, the need to demonstrate that the trait was polymorphous—that it was differentially distributed among the population—was denied. Evidently not all males were seen to commit infanticide, but supporters of the sexual selection hypothesis reiterated their conviction that infanticide was condition-dependent. It was, as Carel van Schaik explained, "a behavior that you do in a very specific set of conditions. And therefore chances are that a male goes through his life and never gets into those conditions, never expresses the trait even if he possesses it." The demand for heritable variation over time, while acceptable as a formal test of evolution, was not considered relevant where selection pressure is as strong as that produced by infanticide. Tim Clutton-Brock emphasized that "if there are things that are terribly important, the presence of a head, the presence of eyesight, then selection may have removed all heritable variation from the population." The proposition that one should measure male lifetime reproductive success to discover whether infant killing was subject to selective pressure was therefore unnecessary—it was not relevant to understanding how natural selection might have influenced the evolution of behavior in the field.

Evolution, Theory, and Data?

In fact, the ultimate explanation for the persistence of the infanticide controversy, offered by researchers on both sides of the debate, was that the opposition had misunderstood the nature of evolution. Every single participant in the controversy declared their commitment to the Darwinian framework of evolutionary biology and either stated outright or heavily implied that opponents had failed to grasp how and why one should go about examining adaptation and natural selection within the field environment. So, for example, supporters of the sexual selection hypothesis drew a distinction between proximate and ultimate factors. Carola Borries stressed that "they argue on the proximate level, and my arguments are ultimate," and Andreas Koenig argued that "it's not important on the ultimate level [what the proximate mechanism is], it's important what the result is."[23] In response, however, Bob Sussman insisted that "they don't understand evolutionary time. They're thinking about evolution with a day-to-day, generation-to-generation perspective. . . . I would think that Darwin . . . would always say, 'I'm not talking about a struggle right in front of your eyes.'" Again, people steeped in sociobiological theory were going to the field with the expectation that they would see evolution occurring rather than realizing that "most populations are basically stable genetically (in Hardy-Weinberg equilibrium), and nothing's happening genetically in those populations each generation." Jim Cheverud suggested that supporters of the sexual selection model might not "understand evolutionary theory . . . in the way that I understand it, as a quantitative science with natural selection, evolution occurring by the natural selection of heritable variation." And while Sussman concurred that distinguishing between ultimate and proximate explanations was fundamental, not surprisingly he weighted their importance rather differently:

> I'm really interested in proximate strategies; that's what I want to study. . . . That's what I think will lead us eventually to good theories about evolutionary mechanisms. But I think what these guys are doing is working on proximate mechanisms and weaving evolutionary theories, and then coming back and trying to prove evolutionary theories with proximate mechanisms, never testing their theories as far as how reliable [they are] in an evolutionary sense. Often data are collected in a weighted manner, what we call stacking the deck—data are collected that fit the theory, but those data not fitting the theory often are ignored.

This orientation can be seen as symptomatic of a particular approach to the relation between theory and data in the field, an approach that might be considered a classic example of the division between deductive and inductive science, a division identified by participants in the controversy themselves. Supporters of the sexual selection hypothesis frequently argued that the opposition to this model was just one example of a dislike of model building in science in general.[24] So, for example, Charles Nunn felt that opponents of sexual selection were "the kinds of people who don't embrace theoretical models or constructs readily or easily." Charles Janson argued that Sussman's critique of infanticide was not "all that different from his general skepticism about generalizations in primatology. And if you look at his [primatological textbooks], it's very clear that his goal in life is to document what is, and not necessarily to worry too much about whether it does or doesn't fall into a pattern." Leslie Digby pointed out that Sussman was not alone in this:

> There are certain authors, scientists out there, who are so meticulous. They may not write many papers, but when they come out with them, everything is beautiful, clear, and very well supported. And then there's that other group. I don't want to name names, but that just [say], "Oh! Got an idea!" and they whip out a paper, or a series of papers, and the data isn't there. But it gets

> people talking, and people say, "No, you're full of it, . . . here's the data that contradicts it." And okay, that idea dies, but two other ideas emerge in the process for which they do have the data that support this. And it's just different philosophies on how you go about it.

Most supporters of the sexual selection model embraced a theoretical, problem-driven, deductive approach to the study of behavior, one that was—as I showed in chapter 2—firmly based on the sociobiological synthesis.

To Sussman and his supporters this approach was fundamentally mistaken. He remembered his early days in the field, before the advent of sociobiology, where "there was such an interest in finding proximate mechanisms and comparative data, and the variations that you find within populations." From his perspective, had primatology continued this early trend of

> comparing two species in the same environment, or one species in two different environments, or three different environments and then trying to figure out how the environment affected [behavior], which would take a long time, but . . . if you did that, by this time, if everybody would have done that and tried to test their theories about ecology, I think now we would have gotten to a much better level, to a set of evolutionarily relevant theories. Because I think now, with most of the data collection, you can predict what a person is going to say before they go out in the field. If you know what his or her theory is, in sociobiology, you know what he or she is going to say. You often know what his or her conclusion will be before the data is collected.

As in their publications and public pronouncements, critics of the sexual selection hypothesis stressed again and again that the data cited to support that hypothesis were inadequate, neither collected nor analyzed appropriately.[25] From their perspective, people were going into the field with blinkers on, and they saw—most of the time—what they expected to see. They were prevented from recognizing that the real world is far more complex and messy than their models intimate.

But for other researchers, not only was theory fundamental to data collection but it was far more mutable than these critics supposed. Louise Barrett made the point that

> you can't just go out and start collecting lots of things, you have to have something in your head if you're going to make progress, you have to struc-

> ture your world somehow. . . . Everything's equally valid if you don't have any theory or any overarching thing, and so you might as well record how many times they pick their nose. . . . Just getting more data doesn't mean you're going to get better theories, because all you can do is hope that the theories you've got will be refined and improved on and approximate the truth, or give you greater knowledge of what the truth is likely to be.

Without a theory to frame your vision of the world, data collection would be impossible: as Robin Dunbar pointed out, "You usually need the theory to recognize the data." Charles Janson argued that "you need theory as a conceptual framework, and that does have the possibility that your framework is wrong, and it could be that for twenty years we will end up leading primatologists down a garden path, which will end up being fruitless, but if so it will be solved." Dunbar went on to note, "It's very clear from the history of science, it seems to me, that what has made a huge difference to the rate of change in certain disciplines has precisely been being able to exploit theory, to provide very precisely testable hypotheses. When you can do that, things move much faster. You may end up by showing that the original theory was wrong, but you get there more quickly, rather than faffing around." Again this was echoed by Volker Sommer, who asserted, "I'm a constructivist. I strongly believe that our theories create our interpretation of what we consider the world. So, in a way, we are making up what we call reality. And, okay, we can make it up in a more or less plausible way. I believe that the way I make it up is more plausible than the way other people make it up, but I would not dream of saying that my way is more true than the other way."

While opponents flatly deny the validity of the hypothesis, at least some supporters are willing to entertain the (unlikely) possibility that they are mistaken. In the context of this question about the appropriate relations between theory and data in the field, it is interesting that contesting primatologists were so willing to accept their opponents' data as valid even as they rejected their interpretations: this willingness may go some way to explain how the controversy could be so prolonged.[26]

Contingent Accounts of Personality and Politics

To a certain extent, researchers did blame personal hostilities for the prolonged and acrimonious nature of the debate. In its initial phases, they recalled that people on different sides would refuse to attend the same conferences or to employ students trained in the "wrong" departments, would set out to sabotage papers and conferences organized by the opposition, would even refuse to speak to one another. Certainly the early stages of the sociobiological revolution, as Phyllis Lee described it, were marked by "a generation of extraordinarily competitive, driven, male researchers": as a result, areas of disagreement were unlikely to be resolved by mutual accommodation. The situation was slightly different when the controversy reemerged in the 1990s: most, though not all, researchers who supported sexual selection emphasized their respect for Sussman's position and the collegial approach they felt he had taken.

Notably, however, they also emphasized just how isolated Sussman was in his position. Kelly Stewart suggested that the resurrection of the debate "shouldn't have made the splash that it did, because it's so few who [disagree]." When pro–sexual selection researchers responded, this gave undue prominence to the views of a tiny minority: "[because] it wasn't ignored, you get the impression that there's a big school out there . . . when actually maybe there isn't." Alexander Harcourt argued that Sussman was "marginalizing himself" within the primatological community, and Leslie Digby wondered whether the reemergence of controversy was just the result of "a small group furthering their agenda." Naturally Sussman emphatically disputed the significance of his minority status, since "consensus does not good science make. Good science is not related to winning the most followers. We must remember that eugenics was very popular among scientists and the majority opinion in elite American and European universities in the early twentieth century." From his perspective, "Most young scientists now, they're not going to try and fight Wrangham and Hrdy and all of those people at the major universities. They're a little afraid of them because they might not get a job. . . . I'm not worried about that. I'm tenured. I've produced a lot of publications. . . . I'm not worried about being called a bad scientist, any of this. And I'm not going to let them intimidate me." So, on the one hand, that relatively few voices opposed the sexual selection hypothesis demonstrates the insignificance of the attempts to continue the debate: on the other, the status and influence of researchers based at elite institutions discourage less consequential people from expressing their concerns about the accuracy of the hypothesis.

Personalities aside, it was clear that the institutional and departmental locations of the people involved, as well as the disciplinary positioning of primatology, particularly in North America, was evidently the most important element accounting for both the persistence and the intractability of the infanticide debate. Again, recall that it was only among primatologists that infanticide remained controversial: other behavioral sciences accepted it as a given. But ever since Sherwood Washburn's establishment of the "new physical anthropology," primatology and primatologists in the United States were to be found within the curricula and departments of anthropology, not in the zoological or psychological niches more common with their European colleagues. The consequences this had for both the infanticide debate and the wider reception of sociobiology within that country were twofold. First, this institutional location meant that the relevance of behavioral primatology to understanding humanity was far more immediate. Primates were a special case: unlike all other animals, they were the business of the discipline that was primarily oriented toward examining the origins and diversities of the human condition.[27] Thelma Rowell remembered the way biologists at Berkeley treated her as an outsider, because "zoology stopped before you got to primates. It was quite extraordinary. Zoology was about mice, basically, and it could be about mountain lions, but it wasn't about primates. Primates belonged to anthropology." As a consequence, as Leslie Digby suggested, primatological research in the United States "typically has to come around to how it helps us better understand humans,"[28] a point echoed by Phyllis Lee, who believed that the reason the infanticide debate had been played out primarily in North America was that "in America, you're trained in an anthropology department, where your take on understanding primates is a take on human nature." Primate behavior, in this tradition, had more immediate consequences for human behavior than did that of other animals.

More subtly, the traditional approach to the disciplinary organization of American anthropology also exacerbated the debate in the United States. Here anthropology had customarily taken what was known as the "four field" approach, where the discipline was organized around the four subfields of cultural anthropology, physical anthropology, linguistics, and archaeology. In practice this meant that anthropology departments, unlike almost any other academic location, habitually contained researchers drawn from both the natural and the social sciences. Straddling this fundamental division in academic life meant that anthropology was particularly vulnerable to the intellectual tremors rumbling through the academy with the debates surrounding sociobiology in the late 1970s and the rise of postmodernism in the late 1980s and 1990s. Both movements created a context in which the interests and convictions of the natural and the social scientists were potentially violently opposed: by the most simplistic of readings, sociobiology threatened to render the social sciences obsolete, while postmodernism denied the validity of the natural sciences. Both movements resulted, according to those I interviewed, in a situation where the cracks already present in anthropology departments widened and deepened into chasms. Almost uniformly, supporters of the sexual selection hypothesis were convinced that infanticide remained controversial in primatology because of the close relation between human and nonhuman primates and because cultural anthropologists and other social scientists felt such a pronounced distaste for biologically based explanations of human behavior. Primatology's location in departments of anthropology, rather than zoology or anatomy, had isolated it from the other sciences of animal behavior on this matter and others.

Anthropomorphism?

To deal first with the close relation between human and nonhuman primates, there was almost uniform accord between supporters of the SSI hypothesis that opposition to sexually selected infanticide was based on a deep reluctance to accept that such a hypothesis might also apply to humans. So, for example, Kelly Stewart said simply, "You don't want to say that humans are adapted to do such hideous stuff as killing babies." Andreas Koenig felt that "people don't care if you're talking about lions or beetles, [but] if this is applied to humans, or to close relatives of humans, nonhuman primates . . . then some people don't like it." Charles Janson echoed this, arguing that "you do have to remember that primatology started out and is still largely a discipline practiced by anthropologists, not biologists. . . . If you see infanticide in an insect, then you're not likely to think, whether it's good or bad, that it applies to you." Carel van Schaik affirmed that "when a rat does something, it's not threatening. When a chimpanzee does something, it's really threatening." And he went on to explain why he felt that at least some of what he considered the fears of the opponents of SSI were realistic responses to the potential consequences for the interpretation of human behavior:

> In America at least, most primatologists come from anthropology. And you study primates because they inform the human condition. And if it turns out that it's adaptive for primate males to go around killing babies then the inference is nigh . . . that, oh, we should see it in humans. As in fact we do. And therefore you get these apologists saying, you can imagine the defense lawyers saying, "Yes, this man killed these babies, but it was only adaptive behavior, blah, blah, blah." And I can see why people get very nervous about that. So you want to be very sure that you are dealing with facts.

On this reading, then, the reason for the hostility to the SSI hypothesis can be found in its implications for human nature: if this act is part of the primate behavioral repertoire, then it may well also form part of the human repertoire.[29] Infanticide is morally repugnant and emotionally distressing, and so opponents of the sexual selection hypothesis commit the "moralistic" and anthropomorphic fallacy, assuming that what should not be therefore cannot be.

These claims that opponents of sexual selection commit the cardinal sin of anthropomorphism were returned with interest.[30] In the first phase of the controversy, Dolhinow in particular had argued that the adaptive hypothesis involved the assumption that langurs had unproven cognitive capacities, particularly in relation to their presumed facility in calculating the likely paternity of any given infant, a point she returned to in the later periods of debate (Dolhinow 1999). Sussman formulated the problem rather differently. He addresses not so much the implications adaptive infanticide would have for human behavior as the fundamental problem of studying in animals a behavior that was first identified and defined in relation to humans.[31] From his perspective, not only anthropomorphic but tautological reasoning is being used here. He emphasizes that the term applied to the behavior under discussion should be not "infanticide" but "infant killing." For him, the legal term infanticide carried a weight of baggage that made its use inappropriate, not least because it could be taken to imply conscious intent on the part of the animal.

The question of intent had, of course, figured largely in the debate, as part 2 showed. That supporters of the sexual selection hypothesis believed the attacks of strange males were confined to infants of a particular age with unfamiliar mothers was a key element in the conclusion that infanticide represented an adaptive strategy. Words such as "targeted," "purposeful," and "intentional" were all used to describe the actions of males in such circumstances—but there is a difference between "intent" and "conscious intent." It is difficult to describe and discuss animal behavior without using "words that we normally use when talking about humans" (Alexander Harcourt), and behavioral researchers have commonly used language in this way, even utilizing metaphor.[32] But there is always an implicit acceptance that this *is* a metaphorical use of language and an understanding that these accounts must be prefaced by the reader with the caveat that the animal was behaving "as if" such decisions or conscious motivations were present.[33] Louise Barrett suggested that since human infanticide is a premeditated act, then "in some ways that gets taken wholesale into the primate world, where in some sense it's seen that infanticide is a choice that the male makes," which is of course not true. As Leslie Digby emphasized, "Natural selection works as long as you end up having more kids (or genes in the next generation) as a result. . . . One of the things you tell an introductory class is that it doesn't have be a conscious decision, as long as they behave this way, and it has this effect, it can be selected for." Carel van Schaik remembered somewhat ruefully "the days when we made up fancy Greek and Latin words for things; there was something to be said for it, because at least people couldn't abuse the terms." Similarly, Robin Dunbar compared the situation with that of particle physics, which

> is littered, in a way that no other discipline except evolutionary biology is littered, with the use of metaphors. All those terms, like the flavors of quarks being up and down, and strangeness and beauty and they don't get confused. . . . I think the problem at the behavioral level that we have is that we are just very good behaviorists ourselves, so we kind of intuitively recognize a lot of these things, and it probably does tend to mean that it's easier to misconstrue what's being referred to. But at the end of the day, in my view . . . it's incumbent on the reader to make sure they understand the background.

From one side, the perception was that there had been a fundamental misunderstanding of the way behavioral researchers speak and write about their subjects. From the other, however, Thelma Rowell pointed out that "if you said, wait a minute, there really isn't any evidence for that, they would say, Oh, you're just a wuss because you're frightened of the idea that your dear animals are doing such a nasty as committing infanticide. . . . And that really shuts you up, because it's a difficult one to answer, because the more you state the case, the more you sound like a bleeding heart." Anthropomorphism, in the sense of ascribing human cognitive and emotional capacities to animals, clearly represented a major resource for scientists in accounting for the debate.

Additionally, however, the identification of the behavioral complex known as "infanticide" could more practically be interpreted as an anthropomorphic example of false analogies, in the sense that a behavior known to occur in humans was stated to occur in animals as well. Then, since the same label was used, they were assumed to have had the same origins. In Bob Sussman's words, this was "one of the worst things they're using infanticide for, human interpretations and back again. First they say, okay, it's a human trait, then they see a superficially similar behavior in an animal, then they compare what they see in the animal to support their interpretation of human behavior, but they actually defined it in humans to begin with." Animal behavior is not only being seen through human eyes but is being inappropriately interpreted according to human experience. While supporters of sexual selection argue that opponents are being inappropriately anthropomorphic in rejecting infanticide because of its implications for human behavior, then, opponents maintain that they have been equally anthropomorphic in failing to distinguish clearly between analogous and homologous behaviors and in their assumptions about the cognitive capacities of the animals in question.

Sociobiology and Postmodernism

The association between primatology and anthropology not only foregrounds the consequences primate behavior is thought to have for human behavior but also provides another resource that supporters of the SSI hypothesis use to explain the persistence of opposition: the antipathy many social scientists feel toward biology. Historically, Robin Dunbar argued that the "problem with anthropology is that it threw away biology at the turn of the century," a rejection reinforced a few decades later by the events of World War II in Europe. Carel van Schaik agreed, pointing out that if "you tried to biologize anthropology, where for pretty darn solid historical reasons anthropologists had turned away from that, you can see how you're shaking the foundations of people's worldviews." They believed this rejection had led the two disciplines to develop with very few positive interactions over the twentieth century. For primatology, this situation then came to a head in the debates surrounding individual and group selection in the late 1960s and early 1970s. There was a lingering perception that group selection had persisted rather longer in primatology than in other areas of behavioral biology. And even where direct support for group selection was lacking, a strong antipathy toward individual competitiveness still was central to the construction of the new sociobiological synthesis.[34] Several supporters of the sexual selection hypothesis positioned themselves as the heirs of this new tradition, in opposition to their more conservative elders who, as Robin Dunbar put it, would have found it "hard to learn a whole new paradigm . . . particularly something as theoretically highly structured as that." On this reading, the initial stages of the infanticide controversy were themselves the early stages of paradigm change for primatology.[35] However, this does not explain the resurgence of the controversy in the 1990s, when both sides proclaimed themselves committed, Darwinian evolutionary biologists.

It may be useful to introduce a distinction between "sociobiology" and "evolutionary biology," since opposition to one does not imply hostility to the other. It is clear, for example, that Sussman's opposition to the sexual selection model is closely connected with his belief that sociobiological theory is used far too uncritically in accounting for behavior. A book he edited in 1999, *The Biological Basis of Human Behavior,* was directly intended to rebut sociobiological analyses of human activities. In his view a "lot of this stuff is pure popularization, without any science, it's just like writing a novel," but since "sociobiology is very close to social Darwinism and can easily be used by neoeugenicists," it presents a rather more insidious danger than pure fiction. In contrast, Sarah Hrdy is emphatic that sociobiology's critics are notable for "how hard they're working not to acknowledge . . . the fact that sociobiology between 1975 and 2002 has changed. . . . The way I'm doing sociobiology today encompasses developmental psychology, history, culture, ethnography—they're all part of what I'm doing, and you don't see me talking about genes for anything, because I know all I can study are phenotypes." While accepting that earlier versions of sociobiology—especially popularized accounts—had been both simplistic and dogmatic in trying to demonstrate the biological basis of behavior, many supporters of the sexual selection hypothesis were also critical of the way opponents still treated these outdated accounts as legitimate targets in the debates. In a sense, as Louise Barrett pointed out, the media's willingness to continue discussing the "gene for" a particular behavior meant that for laypeople,[36] sociobiology still appeared to be a highly reductionist science. In fact, supporters of the sexual selection hypothesis argued almost unanimously that opposition to the hypothesis was the visible tip of a much larger antisociobiology iceberg and was based on a basic misunderstanding of how biological thought and theory had developed since the early 1970s.

Researchers based in the United States went even further. They argued that anthropology had not just rejected biology but was caught up in a wider process through which science and scientific thinking were themselves being rejected by significant sections of the intellectual communities who had come under the pernicious influence of postmodernism.[37] For example, Carola Borries, Andreas Koenig, and Charles Janson, all then based at the State University of New York at Stony Brook, spoke of their shock at discovering that not all their anthropologist colleagues considered themselves scientists and their deep regret at the consequent difficulty of communication.[38] For Carola Borries, it was as if "we're in parallel universes or something," where meaningful discourse was terribly hard to achieve.

Charles Janson argued that the differences in worldviews could be tied to a contrast between the desires to systematize information and the preference for preserving variability; in other words, the search for patterns in what is known versus the unwillingness to give priority to any particular way of interpreting what might or might not be known. While Carel van Schaik acknowledged that useful insights might have emerged from deconstructionism—for example, that it's helpful to be reminded that "facts aren't always as hard as they seem"—he also felt that "[it's] been taken in my view to way too extreme a position, that therefore anything goes and nothing is true and everything is politically motivated and forget about doing science. I'm still firmly a scientist. I still have great reverence for facts and for tests and for the self-correcting nature of science." Sarah Hrdy in fact "wondered whether the science wars [weren't] in fact an outgrowth from antisociobiology rhetoric and positions." Clearly the wider cultural hostility to science that these researchers perceived and characterized under labels such as the science wars, postmodernism, and deconstructionism was felt to be why the "facts" of the infanticide controversy were being ignored: if facts were always subject to interpretation, then it was understandable, although infuriating, that such challenges could be made.

Note, however, that it was against this background of a perceived fissure between the natural and the social sciences, aggravated by the presence of both approaches within departments of anthropology, that the opponents of the hypothesis chose to publish their rejection of the idea not in a primatology or animal behavior journal but in the *American Anthropologist,* the flagship journal of the American Anthropological Association. For supporters of the hypothesis, it seemed that Sussman and his colleagues were deliberately addressing a (relatively) ignorant audience that they thought was predisposed to be critical of both biology and science and were presenting the arguments of a tiny minority of primatologists as if they were representative of a much wider swath of scientific opinion. This audience would have neither the empirical nor the tacit knowledge to critique the article: for example, it would take a solid knowledge of the primatological literature and tacit knowledge of the basic constraints of animal fieldwork in order to realize that Jodhpur's domination of the database was to be expected, given that langurs had been studied there longer and under better conditions than at any other site. For some participants in the debate, this was a deliberately provocative move.

However, as the earlier discussion of the interplay between theory and data in field science showed, at least some of Sussman's opponents were willing to concede at least the bare possibility that his position was justified and to pay tribute to the importance of devout skeptics in the development of scientific debate. So, for example, Carel van Schaik pointed out that the "good thing about the questioning attitude that Bob Sussman and colleagues have, the skeptical attitude he has pushed over the years, is that now people come out and publish their raw observations on it" and that he "and his colleagues, have been more than open about it, have been really, I would say collegial and properly scientific about the whole debate." Robin Dunbar commented that he was "all in favor of people defending hypotheses to the death, because nobody else is going to defend it, so in the end, in the interests of science, somebody has to be prepared to put their neck on the line." From this perspective, science can still proceed through disputation and refutation, even where a proposition has been "affirmed to such a high degree of certainty that it would be perverse to withhold one's provisional assent."[39]

Conclusion

At the beginning of the twenty-first century, the positions adopted by participants in the infanticide debates can be roughly summarized as follows. Supporters of the adaptive model took it as axiomatic that infanticide occurs as a result of sexual selection and is a strategy primate males use to maximize their individual reproductive success. They acknowledged how rarely complete observations of infanticide are made and pointed to the immediate context of these events: infant deaths and disappearances occur when new and nonrelated males have the opportunity to breed with their mothers. They tend to be followed by the male's mating with the mother, who ordinarily produces the next infant sooner than had her previous infant survived. The words "usually" and "ordinarily" are important here: supporters of the hypothesis would not expect to see infanticide every time a new male enters the group, nor would they argue that all infant deaths are caused by marauding males. It is a condition-dependent behavior, one that is in all probability "fixed" in those populations where it is seen. Finally, it is also a behavior likely to have had a profound effect on the evolution of primate social systems in general.[40]

In contrast, opponents foregrounded the lack of detailed observations of infant killing, just as earlier participants in the debate did, and cited the prevailing use of assumption, theory, and postulation in the infanticide literature. Perhaps provoked by the increasingly significant claims about the importance of infanticide in the study of primate behavior, they did not stop there, as their predecessors did, but went on to lay out the steps one must take before one could suggest that infant killing might have had any evolutionary effect at all on primate society. First, one must demonstrate the genetic basis of the behavior, and then one must show that there is a difference in fitness between males that possess this trait and males that do not, in that males that do possess it should leave more offspring. Only at that point can one begin to speculate on the consequences of this behavior for primate social systems, and in doing so one must also take into account the wider context of the infant killing event, in terms of the total pattern of infant morbidity, genetic relatedness not just within groups but within the local population, and the more general levels of aggression shown for different groups and species in different social and environmental contexts.

Leaving aside for the moment the contingent issues researchers raised in explaining the longevity of the infanticide debates, it is clear that each side also believes the other is making unreasonable claims. For the opposition, it is unreasonable to use the *absence* of an event (a failure to observe infanticide in a given population) as an indication of the importance of infanticide risk in the evolution of relatively permanent male-female relationships. For the supporters, it is unreasonable to expect that a "gene for" infanticide must be produced before the consequences of this behavior can be explored in relation to the data available. However, considered in relation to the broader frameworks that researchers have adopted when approaching the problems and advantages of carrying out science in the field, each position is both understandable and consistent. As Hrdy and her colleagues noted in their reply to Sussman, Cheverud, and Bartlett's 1995 piece in *Evolutionary Anthropology*, there is a profound stylistic and practical difference in the way different researchers approach the conduct of science, especially in the field: whereas some are absorbed by observing and recording data, others find fascination in slotting that data into broader patterns of perception. This does not mean that some are obsessed with detail while others abandon data for flights of fancy, but it does mean that some are frustrated by the unwillingness of others to recognize similarities in events across sites, species, and taxa, while some fear the consequences of generalizing in advance of one's data—especially when addressing the potential biological basis of human behavior.

One question that many supporters of the hypothesis repeatedly asked was, Why infanticide? Given that the objections their opponents made to the data used to support the sexual selection hypothesis could be applied to many other examples of primatological theory and practice, why had they chosen to focus on infanticide in particular? The answer, naturally enough, does come from the contingent repertoire: for the human observers, infant killing is a revolting and reprehensible act. Infants are there to be protected, not to be attacked and injured. Obviously it is inappropriate to judge animal behavior by human standards—but when primates do things, there inevitably are consequences not necessarily for understanding human behavior, but certainly for its interpretation. Primatologists know this: historically and at present, it has been a theme in their requests for financial funding and in the omnipresent media coverage of primatological research. In the United States—where the controversy has been most at issue—this is even more pressing, given that primatology tends to be institutionally situated within anthropology departments. But the infanticide controversy cannot be explained solely by referring to the moralistic fallacy, not when participants in the debates are themselves practicing and respected scientists. Opposition to the hypothesis was not based purely in anthropomorphism, and it did not arise from opposition to evolutionary theory: it sprang from the solutions that primatologists had found to the "fieldworker's regress," the questions raised about the appropriate theoretical frameworks and methodologies for conducting science in the field, and the (for the most part) mutually tolerated persistence within this discipline of different answers to the same questions of practice, as the next chapter will show.

SEVEN

Controversy and Authority, Narrative and Testimony

Introduction

To understand why the subject of infanticide initially became controversial for primatology and why it has proved so difficult for primatologists to resolve the debates, we must return to the questions and issues first raised in chapters 1 and 2: the circumstances in which behavioral primatology emerged as a field-based scientific discipline. In the late 1950s and early 1960s, those interested in primate behavior faced a series of interrelated problems: how to create a scientific tradition for the field, how to manage the expectations and assumptions of coworkers drawn from disparate disciplines, and how to deal with reports of primate behavior that could conflict with each other. If one researcher reported that a species at one site behaved in a way not seen at another, did that mean someone had made a mistake? Were different ecological conditions eliciting different behavioral patterns, or were they affecting the observers' ability to see the animals? Or were members of the same species in fact behaving differently in different places? If so, what were the consequences for primatology's often-declared mission to study the nonhuman primates so as to better understand human behavior? These problems of field practice, which I will argue constitute the "fieldworker's regress," and the proposed solutions, both explicit and tacit, not only played a fundamental part in the emergence of the infanticide controversy and its perpetuation within primatology but also were responsible for *preventing* its appearance until sociobiology emerged in the early 1970s.[1]

The emergence of this new sociobiological synthesis was clearly responsible for introducing dramatic changes in the way researchers conceptualized and analyzed the evolution of behavior. However, we must not forget that, at the same time, major changes were occurring in field-based scientific practice. By this point, as part 1 demonstrated, systematic studies of wild primate behavior had been in progress for almost a decade and a half. While much of the information gained from these studies was based on relatively brief periods of field research, the value and richness of long-term behavioral data was increasingly being recognized, and researchers were making efforts to ensure that they could keep these long-term research sites going and growing. Even as the histories of the primates at these sites were being written, so too were the numbers and geographical range of primate field sites expanding. By the early 1970s they covered not just the large-bodied terrestrial primates of Africa and Asia but, increasingly, the animals of South America and the nocturnal primates.[2] And at these sites the ad libitum collection of data was being steadily replaced by the systematic attention to choice of sampling methods that Jeanne Altmann had called for in her 1974 paper, widely circulated before publication (Haraway 1990), making it *theoretically* possible for researchers to make direct comparisons between primate behaviors at different sites. In acknowledgment of the importance of long-term data collection, more and more local people were employed by foreign scientists not just as guides but as field assistants, and the need to build relationships between foreign researchers and local universities—especially in South America—was also recognized. Primatologists and primatological work were becoming much more tightly tied in to local and international networks of cooperation, alliance, and competition. In particular, the stress on observing behavior at "undisturbed" locations was diminishing, partly since no site can realistically be described as "undisturbed" in an industrializing and globalizing world, but also in response to the endangered status of many primates and the need to establish how human involvement with local and global ecosystems was affecting them. The development of sociobiological thought developed alongside these changes in field experience: theory and practice evolved hand in hand.[3]

In this chapter I will explore the ways that the infanticide controversy was influenced by the evolution of primatological field practice and by the new strategies and expectations that both complemented and extended those characteristic of the first phases of studying primates in the field. I will show why the subject, though shocking, at first generated comparatively little controversy and why—in relation to the 1960s standards of field methodology and reporting—it was relatively easy to challenge the accuracy and reliability of Hrdy's early articles. I will examine Hrdy's approach to the Abu data and the other langur field studies in relation to the wider changes in field practice that were occurring in the mid-1970s and in the context of the strategies authors used to ensure that other researchers treated their reports as veridical accounts of events in the field. I will also discuss the potential consequences of challenging this veridicality. Although it is a truism, field results cannot be replicated. No researcher would realistically propose spending years studying the same population to reconfirm or repudiate claims made about its behavior. Not only would that be practically and pragmatically pointless, it would be impossible: the passage of time itself would have changed the individuals, their relationships (both genetic and social), and the ecological context of their existence (Rees 2006a).

This chapter will show that the fieldworker's regress framed the evolution of the infanticide debates in primatology. Recall that in the introduction I counterposed this to Collins's notion of the "experimenter's regress"—the idea that despite the significance attached to the concept of "replication" in traditional accounts of the scientific method, in practice one cannot successfully replicate an experiment in situations of controversy. Collins showed that where extraordinary claims were being made or novel procedures undertaken, it was always possible to challenge an opponent's experimental procedure in order to explain why the "wrong" result had been achieved. The unstated presumption, of course, is that if the "right" procedure had been followed, then the right result would ensue, since science depends on universal laws and laboratories are spaces specifically constructed to produce controlled, predictable, and universal results. In the field, things are different. Field sites are unpredictable, chaotic places, not subject to controlled conditions and often sought out precisely for their unique properties. To an extent, the spaces can be treated as functionally similar, in the sense that just as for the experimental sciences the question is, "Is the apparatus the correct way to test the theory?" so for the observational sciences the question can be, "Is the site likely to elicit the appropriate behavior or phenomenon?"[4]

However, the difference lies in one key fact that became evident as fieldwork in general—not just in primatology—came to maturity through the 1970s and to the present. This is the degree of interpretive flexibility that is habitually required of the researcher in answering the second question posed above. Unlike the lab scientist, the fieldworker is consistently confronted with what I have called the fieldworker's regress. Contingent questions were recognized and discussed as potential problems from the inception of fieldwork: data collection and the standards for doing it evolved hand in hand. Not only was the "observer effect" recognized and explicitly interrogated from the outset, but even the effect the observer's personality might have on the results of the study was regarded as a legitimate topic for discussion. Whereas in the laboratory the experimenter could be expected to know and to account for every variable involved in the experiment, in the field the observer's knowledge was inevitably partial, with scientists frequently having to invoke unseen events or forces to explain or to critique their own research. Fieldworkers, in short, had to learn to be reflexive about their research and to come to terms not just with the physical discomfort of life in the field but with the intellectual uncertainty the field environment imposes. The experimenter's regress is evoked in situations of controversy because the lab space—and naive forms of philosophy of science—are structured to evoke expectations of dependable results. If the results seem uncertain, then something has gone wrong. The fieldworker's regress, on the other hand, is the default state for the field scientist—a constant state of interpretive flexibility perpetually provoked by the structure of the field space.

The Story So Far

Returning then to the infanticide controversy within primatology, it is evident that the signs that would assure the wider intellectual community that the author of a report was worthy of trust were also changing rapidly over the latter half of the twentieth century. Initially the descriptions of primate behavior that existed early in the century were considered untrustworthy and inaccurate by the nascent behavioral biology community. At the beginning of the century, recall, key figures in the biological and behavioral sciences had realized that if the evolutionary origins of animal behavior were to be understood, that behavior would have to be studied in conditions that resembled, as far as possible, the conditions under which it evolved. By the end of the World War II, numerous researchers in animal behavior were eager to emphasize the importance of studying animal society (Scott 1950), not least to make that information available to the politicians and social scientists who were trying to deal with the large-scale social problems that beset the urbanized communities of Western society. In this the study of apes and monkeys was to play a special part. However, the accounts of primate behavior then available were still sporadic and scattered. While some reports did appear in journals sponsored by European and North American scientific societies, most were published in the autobiographies and narratives of hunters, travelers, and missionaries who had frequented the areas where wild-living primates were to be seen. For the most part, descriptions of primate behavior in the wild formed an adjunct to accounts of adventure and survival in faraway jungles and forests, while descriptions of primates in captivity still largely focused on psychological and anatomical characteristics. If useful knowledge about the evolution of primate behavior was to be developed, then descriptions urgently needed to move away from anecdote and toward acceptable data, to reject hearsay in favor of scientific reporting.

To achieve this, a number of pressing practical problems had to be addressed. First, prospective primatological researchers needed to identify appropriate sites to work at and to decide on appropriate methods for the study of behavior there. C. R. Carpenter played a key role here. His description of an "ideal" field report, written in the mid-1960s, advised that the personal qualifications of the observer should always be taken into consideration. It called for researchers to watch as many animals as possible for as long as possible, at a site where animal behavior could be considered "representative" of that species and where human contact was kept to an absolute minimum. The field reports that followed did their best to attend to these injunctions, but since the ideal length of a field study was then unknown,[5] and since the number of animals observed depended on the number known to be present at the site,[6] these factors tended to be defined in terms of what was practically possible, rather than what was ideally desirable. Similarly, while researchers did their best to select sites that were undisturbed and representative, by definition a primate population that was known to exist must have had some human contact. Ascertaining what was "representative" of primate behavior was of course *exactly* what these early studies were trying to establish. Many of the reports made in the early days of field site primatology struggled with these questions as they did their best to reach conclusions that would be acceptable to their peers.[7]

Researchers were, in effect, making a series of practical compromises between what *should* and what *could* be done in a field setting. Evidently, doing field science required a rather different kind of scientific practice than had grown up for the laboratory. While some forms of social experiment were used in the early years of the postwar expansion in primate field studies, their use was limited—the danger of damaging or destroying the primate societies under observation was far greater than the value of the information that might be obtained. Closer attention to the choice and use of sampling methods was used as an alternative means of studying behavior in a controlled and authoritative manner. Habituation and individual identification—where possible[8]—came to be widely used as a means of implementing these sampling techniques systematically. Although experiment and manipulation were excluded, it is clear that the authority of the field scientist was coming to depend on conceptual alternatives to the methods that characterized the laboratory—in particular, the beginnings of a comparative framework and the exploitation of "natural experiments."

These methodological debates were not taking place in a vacuum. Data on primate behavior and relationships were being collected at the same time as these compromises and alternatives were being negotiated, and as I have shown, these data were sought not only out of intrinsic interest but for what they might imply for the understanding of human development. Among other things, researchers studied these most social of mammals to discover the origins of primate sociality, to understand why male and female primates had come to socialize throughout the year, to understand the roles that males played in the groups they settled in, and to determine the functions that group living played for primates at present. This last point was particularly important, since even as the ideas and concepts that were to underpin the new sociobiological synthesis began to circulate and to gain ground among behavioral scientists, there persisted the older notions of group selection and the perception that for primates the group was at least as important as the individual. This occurred not least because, for so many primates, group living seemed fundamental to individual survival and reproduction, no matter what environment they might inhabit. Baboons on the savanna grouped together to repulse predators more effectively and exploit the environment more efficiently. Forest-dwelling langurs depended on the group's ability to socialize the individual into appropriate langur behavior. And there was an evident assumption—at this early stage—that the basic parameters of behavior were determined by species: the classic example was the contrast between langur and baboon behavior.

The way researchers wrote about their animals as primate fieldwork began both reflected and reinforced these assumptions and discoveries. In particular, the use of narrative structures to organize fieldwork accounts not only lent weight to the presumption that there was such a thing as "species-typical" behavior but was also an inevitable consequence of the then-current state of primatological studies. The use of such narratives was relatively common in the first stages of primatological fieldwork, since field studies had just begun and records of primate behavior and development were necessarily patchy and disjointed, exacerbated by the particular characteristics of the primate life cycle. Charles Janson pointed out that, for most animals, "If you study them, say, for the length of a PhD thesis, [then even] if you don't have their entire life span, you've got a good chunk of it, at least half. Whereas for most primates, study them for four years, my god, they've barely grown up! And you haven't even begun to study a tiny fraction of what happens during their adult lives, or seen what happens during the full ontogeny of an individual."

From the late 1950s on, primatologists were encountering strange troops consisting of adults of unknown age plus infants and juveniles that often could not be reliably sexed. It took time for individuals to be identified, and it took far longer for them to be reliably aged. As a result, when reports of behavior were written up, they were not and could not be literal accounts of the growth of an individual from infancy to adulthood, nor could they account for the evolution of troop social relations. Instead, researchers used narrative to weave their observations of many individuals into an ideal type of development.

This strategy was used consistently throughout this early period, as shown in the accounts in DeVore's (1965) edited collection. Reports first described the physical and social characteristics of the research site, then normally moved to the physical and social characteristics of both sexes of animals as the life cycle developed.[9] Beginning with the unsexed infant and its reception by other members of its group, researchers would move through the description of the development of sexed, if not sexual, behavior as infants matured into male and female juveniles, then would follow the pattern of both sexes separately as they became fully mature males and females. This is a practical strategy that makes it easier both to organize the observations of animals at different stages of the life cycle so as to describe them effectively and to present them so an audience can make useful sense of the results. However, it is also a technique that accentuates the authority of the observer and imposes an apparent coherence on reality that may be far more fluid than a naive audience might suppose.[10] The audience was presented with an account of how "the" baboon, or langur, or macaque developed, and of the key elements encouraging the social group to cohere, in a way that both reflected the tendency of classical ethology to seek out "species-typical" behavior and depended on rhetorically absorbing observations of individual animal development into an account that aimed toward universality.

The difficulty was that the answers to the questions about the nature of primate society and social structure were not necessarily consistent from site to site. At some places it seemed that males acted as the defenders and leaders of the group: at others, males were seen to flee from danger, leaving vulnerable infants and females to shift for themselves. Some primate species, such as baboons and macaques, showed behaviors that might be characterized as paternal care, whereas males of other species ignored infants altogether. Several researchers saw no evidence for territorial behavior, and where others did argue for territoriality, some deduced it from the absence of intergroup aggression alongside loud calling, while others recorded knock-down, drag-out, fights between the males of different groups.

Primates were seen to live in groups with single males and with multiple males; in most if not all societies, males were known to form dominance hierarchies, but such hierarchies were not always linear. Some females protected their infants while others were happy to let other troop females handle and examine them. The developing picture of primate sociality in the late 1960s was not necessarily consistent from one site to the next.

However, if primatologists were to transform their discipline into more of a science rather than an example of natural history, they had to do more than simply record the observations made at their sites in a systematic and orderly fashion—they had to be able to make generalizations about primate behavior. Science depends on generalizing—on identifying patterns of relationships, explaining them, and using them to predict what might occur under given conditions. Primatologists working in the field had to find some way to explain the staggering variety of primate experience and behavior by identifying the possible causes of such variation and by showing that the *causes* of primate performances remained the same even where the *effects* were demonstrably different. This is important, since it represents the first realization of the fieldworker's regress: getting different results did not mean error. Rather, it meant that these results had to be situated more firmly in their context of production, in the *place* where they had been recorded. In the first iterations of such explanatory frameworks, considerable attention was paid to the internal characteristics of the group, as researchers sought to account for social order by identifying social roles existing independent of the current occupant. This perspective, deriving from the pioneering work of C. R. Carpenter, took an increasingly anthropological turn in North America and was associated closely with the work of Sherwood Washburn, Jane Lancaster, and Phyllis Dolhinow. On the other side of the Atlantic, where primatology was flourishing in departments of zoology and psychology rather than anthropology, an approach derived from ornithological research was coming to dominance. This approach focused on the particular environment of a primate group as a fundamental factor in social life.[11] Ecology in this sense did not determine society but did exercise considerable influence over the basic characteristics of social life as it was expressed in that place and at that time.

Crook and Gartlan's influential 1966 paper in *Nature* was one of the earliest attempts to bring order to the accumulating accounts of Old World primate variety by relating social organization to species ecology. It revealed a progressive and adaptive shift from the insectivorous, forest-dwelling solitary animals that were thought to resemble the ancestral primate form most closely to the vegetarian, troop-living primates found at the forest edge and on the savanna. Key ecological variables were diet, the availability of sleeping sites, and the danger of predation: social groups of fruit-eating primates in the forest were small because of the limits on their food supply, while group sizes rose on the savanna (food distribution permitting) because the increased predation risk in open country called for constant vigilance. These factors influenced not only the size of the group but the relations within it. As group numbers rose, so did competition for access to females, leading to males' increased size and aggressiveness, which in turn led to increased complexity of group relationships as males developed ways to cooperate with each other. Where food supply was less abundant, however, maintaining several large males within the group was difficult: in such circumstances, one-male groups represented a more efficient means of propagating the species. Of course, the exclusion of most males from mating groups meant that intermale competition in these ecologies was again intensified. Overall, Crook and Gartlan were able to use their three key ecological variables to account for much of what was then known about group size, group dispersion, group social structure, and individual socialization among the Old World monkeys.[12]

The Initial Observations of Infanticide

Not surprisingly, then, the fact that the behavior and social structure of the langurs of Dharwar and Jodhpur differed so sharply from those of the langurs of Orcha and Kaukori was initially explained by the variation in environmental conditions at the sites. As part of their critical elaboration of Crook and Gartlan's arguments about the relation between ecology and social structure, Eisenberg, Muckenhirn, and Rudran (1972) agreed that violence and infant mortality were far more likely under "conditions of high population density or in marginal habitat" (868). Alison Jolly directly related the behavioral flexibility of langurs—demonstrated by the variance in social behaviors seen at different sites—to their ecological flexibility. Unlike other leaf-eating monkeys, langurs are partially terrestrial and therefore can use a wider range of environments—in particular, the edges of the forests and the fields. Living on the margins in this way seemed to result in the higher population densities seen at Dharwar, again intensifying male-male competition and producing the single-male troops and periodic male replacements accompanied by disappearance of infants and juveniles that were seen there. In the absence of such intense population pressure, males born into the troop could reach maturity, producing the multimale troops Phyllis Jay recorded. Finally, Yoshiba's direct comparison of conditions at

Dharwar, Orcha, and Kaukori had concluded that the differences between the sites "can be more easily understood as the result of different degrees of emphasis on patterns common to all the localities" (1968, 241). The biggest difference between the sites was the population density at Dharwar—which had led to the rapid social change and associated infant mortality at that site.

Even though the subject of infant killing retained all the moral and emotional burdens that were to weigh so heavily only a few years later, even though it resided at the nexus of a number of topics (sex, aggression, male-female relationships, and infant care) that had clear political implications likely to inspire strong views in both experts and laypeople, infant killing did not at this stage become controversial. Partly this was because the sociobiological ingredient had not yet been added to this heady brew—but it was also because the establishing consensus on the nature and reporting of field methodology and the then current theoretical conceptions of the foundations of social behavior came together to prevent the emergence of controversy. In the first place, the call for a detailed account of the context of observations meant it was possible to explain differences in behavior by referring precisely to that physical and ecological context. In the second place, the focus of the reports from Dharwar and Jodhpur was on the process of social change and its relation to the creation and maintenance of group structure, *not* on the occurrence and consequences of the act of infant biting itself. As chapter 3 demonstrated, a number of explanations for infant biting had already been suggested, most depending either on mechanisms of group selection that had not been shown to exist (the regulation of population size and the maintenance of genetic diversity within it) or on the immediate *proximate* context and results of the act itself (the speedy formation of social and sexual ties between unfamiliar males and females. At least for Sugiyama, the focus of scientific attention was, and ought to be, on how the one-male pattern of troop organization emerged and was perpetuated at Dharwar: infant biting was only part of this process.

Sarah Hrdy began her work at Abu at a time when the new sociobiological concepts and hypotheses were being constructed. She published her first description of langur behavior the year before E. O. Wilson's synthesis of these ideas appeared, an overview that concluded with the (in)famous chapter on the sociobiology of human society. The immediate and immense political and intellectual consequences of this book are well known, and its reverberations are still felt in the academy. But sociobiology was not simply an attempt by natural scientists to invade the territory of the social sciences and humanities.[13] The concepts of kin selection, inclusive fitness, and reciprocal altruism had a major impact on the way behavioral biologists, among others, went about their work. For the most part primatologists welcomed this, but as I said in the introduction, some not only felt that putting sociobiological ideas to work in the field was impractical, but believed the sociobiological revolution itself might hinder the study of primate behavior in particular. There were immediate practical problems with applying sociobiological concepts in the field. Testing sociobiological models required long-term data on animal kinship as well as on the distribution of local resources, and such data were still being accumulated. Some primatologists were concerned that sociobiology was not just changing how researchers thought about primate behavior but altering researchers' choice of what behaviors to study—a danger where the genetic basis of behavior was the focus of speculation but in practice was almost entirely unknown and where comparative databases were still being constructed.

Hrdy's account of events at Abu was an immediate and powerful demonstration of sociobiological theory's capacity to make sense of apparently aberrant behaviors. But for those wary of this new framework, looking closely at the context in which these behaviors had been observed revealed telling gaps in the narratives she had produced. She had not personally witnessed any of the infant killings that her interpretation of langur behavior sought to explain. Her periods in the field had been short and separated by long intervals of absence. She worked at a site where the langurs lived in close commensal contact with the local human population, from whom she had acquired much of the information essential to the support of her model. She had sought to produce an all-encompassing analysis of langur behavior that turned on a rarely observed event thought to occur at only three places over the vast range the langurs occupied—and three places characterized by very unusual ecological conditions at that. According to the emerging consensus on field methodology in the 1960s, Hrdy's account was open to question: recall Carpenter's demand that the observer's qualifications be questioned; that the animals watched should live in undisturbed, representative groups; and that field reports should in the first instance be descriptive rather than interpretive. Hrdy's work at Abu could be criticized on any of these grounds.

As chapter 4 showed, she had made extensive use of the narrative structure in putting her case for the sexual selection model. In the primatological literature in general, this "ideal type" model of primate growth and behavior was being gradually supplanted by statistical analysis of the records of individual development, records that simply had not been available to earlier writers. Narrative reconstruction, however, continued to be used to describe unusual events—*especially* those relating to social change. Hrdy used a modified version of the technique to account for events at Abu during her absences; she tied "snapshot" observations she herself and local informants had made into a logically coherent chain of linked events that could be read (whether intentionally or not) as a sequence watched in its entirety. She used another version in her *American Scientist* article when she pulled the widely dispersed and somewhat disparate descriptions of langur social behavior into the cyclical narrative described in chapter 4. In this case she added to the rhetorical power of the technique by extending the metaphor of the individual life cycle to the troop itself, and by extension to the species—tremendously increasing the scope and significance of her work but also perhaps taking a step too far along a walkway supported not by solid observation but by a scaffolding of reconstructed inference. Certainly Curtin, Dolhinow, and Boggess took her firmly to task for including so many assumptions and presumptions in her models—but they failed to acknowledge that Hrdy was using techniques tacitly accepted by fieldworkers in the primatological community, techniques that they themselves had used to make sense of their own observations.

By the late 1970s, fieldwork standards and fieldwork practice were shifting, both as a result of the influence of sociobiological theory and as a consequence of the practical problems researchers had faced and overcome. Hrdy's work exemplified this shift. While recording the context of field observations remained central to the practice and the authority of fieldwork, it had now been joined by other techniques. Experimental manipulation still remained largely unacceptable, but as more and more field sites were established and several could now realistically be described as *long-term* sites, real functional alternatives to explicit manipulative testing began to emerge. At the same time, the call for research at undisturbed sites was steadily less pressing as primatologists realized not only that such places represented a research ideal that was less and less achievable, but that research at disturbed sites was extremely important in understanding primate behavior. Finally, local field assistants were becoming more and more essential to the functioning of long-term field sites in particular: Western researchers periodically had to return to their home institutions, but local students and assistants—and sometimes scientists—could maintain the long-term data collection in their absence (Rees 2006a). Seen against this shifting pattern, Hrdy's work at Abu looks decidedly less insecure and far more prescient: her strategies, while rarely used at that time, were to become both more common and more systematized.

Field Science: Biology, Ecology, and History

The question of *where* to observe langur behavior in order to truly understand the basis of langur society was absolutely fundamental to the debate. Once Hrdy had made her sociobiological suggestions, Dharwar, Jodhpur, and Abu were increasingly characterized as unsuitable, highly unnatural places to watch primates. But making such a judgment call was highly risky in itself, as the fieldworker's regress would lead one to expect. The accusation could always be returned with interest, as Hrdy (1978) did when she suggested that behavior recorded at the very margins of the langurs' range (the Nepalese studies undertaken by Dolhinow's students) was unlikely to be "representative" of behavior throughout the rest of the range. The problem was simple: a definition of a "representative" study site simply did not exist, other than the injunction to avoid sites where human influence might have distorted primate behavior.[14] Producing such a definition would be a task akin to cleaning the Augean stables, and the ensuing controversy might have overshadowed the one surrounding infanticide itself. Tacitly, primatologists accepted a researcher's word as to whether a site was "disturbed"; but raising the question was an important strategy in controversial circumstances, leading the audience to question not the reality, but the *significance* of the behavior.

To a certain extent, primatologists still acknowledge that the question whether one should work at disturbed or undisturbed sites is relevant. Louise Barrett drew attention to the practical problems of finding and getting to a "pristine" site, arguing that "your ideal would be to parachute into somewhere and never see another soul, but that's never going to happen." But increasingly, primatologists have come to realize that disturbed sites can still answer interesting questions. So, for example, Alexander Harcourt suggested that though a request to study the behavioral ecology of rhesus monkeys in Calcutta might be refused funding, there were still "plenty of questions that one can ask about rhesus monkeys in the streets of Calcutta. I suspect the objection would be wrong and the person studying the rhesus monkeys would be asking questions that you *could* study in the streets of Calcutta."

Generally, the early assumption that you can study adaptation only in a habitat resembling the one the species evolved in seems to have become less common. Instead, primatologists were now seeking out sites where they could answer the specific questions that had brought them to the field. For this reason, sites that earlier researchers might have dismissed as "disturbed" are becoming valuable, since they may provide access to natural experiments where behavior has been altered, and where adaptation can be addressed and studied directly as animals respond (or not) to the new conditions. Second, as the full range of primate behavioral variability became obvious, the significance of primate adaptability was also becoming clear. As Robin Dunbar pointed out,

> If we've learned anything in the last half century of primate field studies, it's been how adaptively flexible primates are. . . . Their capacity to fine-tune their behavior to circumstances is considerable, and in the end, that's what they have a brain for, that's been their whole evolutionary strategy, if you like. Now what that means in effect is, okay, if you have a disturbed environment, it is a challenge . . . but most primates are sufficiently adaptable that they can cope with a very wide range of ecological circumstances. And those interferences, habitat interferences from humans as it were, in effect, become much more like a natural experiment.

After all, even by the late 1970s, and certainly by the 1990s, "there was enough very basic descriptive natural history material to start asking the larger correlational questions" (Charles Janson). By this point the database of descriptive observations of species behavior was large enough that tentative model building could begin: adding yet another example of within-species behavioral variety without attempting to relate it to what had been seen elsewhere could easily seem repetitive rather than relevant. A new framework for approaching fieldwork was developing.

Sociobiology had by no means replaced socioecology—but it had shifted the focus and had changed the working patterns of most behavioral primatologists. Since sociobiologically based hypotheses could realistically be tested only against long-term data sets, collecting such data in detail remained a significant task for field site workers: ecological information about the distribution of food, water, sleeping sites, predator activity, rainfall, and so on continued to form an essential background to the interpretation of primate behavior.[15] But it was now accompanied by a sharper focus on recording reproductive behavior at the site itself and an awareness that an animal's "decisions" on how to use these resources depended on its individual position: the animal's sex and age and the number and position of its relatives all were now relevant factors that needed to be considered and recorded. Primates were no longer interchangeable examples of age-sex classes but could be treated as individuals with histories of their own, histories it was the business of researchers at the long-term sites to record. Although recognizing and habituating individual primates had been a strategy for studying behavior from the inception of the postwar wave of primate field studies, it grew deeper and more nuanced the longer it persisted (Rees 2007).

Similarly, the second of the key developments in primatological field practice in the 1970s had been recognized at least since the late 1950s, but it required accumulating a certain amount of behavioral knowledge before it could be put into practice. This was the presumption that it should be possible to treat primate field sites as individual iterations within a wider social and intellectual network. In other words, rather than examining the behavior of primates at a given field site in isolation, researchers should be able to place that behavior directly in the comparative context of behaviors recorded at other times and places. Fundamentally, this would provide field primatologists with a realistic and functional alternative to experimentation. Individuals studying two, or perhaps three, primate groups at their site could not risk interfering with their behavior. Doing so not only would irrevocably alter the behavior of that group, thus potentially wasting time spent habituating and identifying the animals, but would produce no statistically significant results. However, if one could gather the records of behavior at many field sites, together with a reasonably detailed account of the local ecological variables, it ought to be possible to test variations in these data. Just as in the laboratory, one could examine behavior in the presence or absence of a particular condition (high population density, a predominantly fruit-based diet, no predator activity). The difference was that the primatologists could not manipulate these conditions themselves but had to wait for them to occur in the ordinary course of events.[16] Of course this strategy raised further questions. First, it took time to put it into operation, since the relevant records had to be made before they could be examined. But second, there was still the question of who the observers were, what they were recording, and therefore whether the same kind of behavior was in truth being observed at these very different sites. This was the kind of question Jeanne Altmann had tried to address in her 1974 paper on the choice of sampling methods, where she emphasized the dangers inherent in a too swift rush to quantify behavior in the absence of careful attention to the way data were being collected.

Both these techniques—the construction of long-term field sites and their placement within a steadily growing comparative network that provided primatologists with a knowledge base that was geographically widespread and historically grounded—could trace their ancestry to the early days of primatological work. However, and critically for the development of the infanticide controversy, the third and fourth changes seen in primatological field strategies during the late 1970s and early 1980s were less rooted in past practice and led to a different kind of field framework. The appropriate use of the comparative literature in particular became a topic of discussion. Sociobiology treated all animals—even humanity—equally, as all subject to the same relentless evolutionary laws. This meant that for sociobiologists the comparative literature on the behavior of other animals was relevant to the interpretation of primate behavior. This was a problem for some researchers, particularly in the United States, where primatology was firmly based in the anthropological rather than the ethnological tradition. For them the comparative literature simply wasn't pertinent: not only that, but making generalizations about animal behavior not just across species but also across taxa was risky. For sociobiologically inclined primatologists, the logical step following the cross-site comparative analysis of primate behavior was to put that behavior in the wider context of other animals. This strategy was evident from the first infanticide conference in 1982, to which specialists in all areas of animal behavior—including laboratory behavior—were invited. It took some time for it to become successful, however; for example, not until Van Schaik and Janson's *Infanticide by Males and Its Implications,* published in 2000, was the primate infanticide literature treated as a completely integral part of the literature on infanticide in larger animals. Primates, again, had habitually been isolated from the rest of the animal kingdom—especially, it seemed, in the United States.

The final strategy that appeared in the 1970s was at once more direct and more nebulous. Put simply, after the advent of sociobiological theories and concepts, people went to the field intending to test specific questions. This was a major change from the situation up to the middle 1970s, where the concern had been—as Carpenter had advised—to record as much behavior as possible for as long as possible. People traveled to field sites to investigate particular kinds of behavior, and they tried to choose their sites according to their estimate of whether the ecological conditions there were liable to elicit the kind of behaviors they wanted to work on. This is not to imply that the theoretical analysis of behavior and the development of explanatory hypotheses did not exist before the sociobiological revolution—indeed, there were primatologists like Stuart Altmann who defined themselves as sociobiologists many years before the "new synthesis" took center stage in academic life—but there was a qualitative change in the questions being asked and the level of behavior they applied to. Rather than focusing on detailed descriptions of the group's behavior, the relationships within it, and the nature of its interaction with the local ecology, investigators concerned themselves with the behavior and interests of individuals. The emergence of long-term field study sites made this possible, in that animals were known as individuals and their unique histories were recounted in site records, and sociobiological theory made these records meaningful. It allowed researchers to ask tightly focused questions about the reproductive consequences of individual behavioral "decisions," often phrased in terms of a "cost-benefit" metaphor that reflected both the importing of game theory into evolutionary biology and the increasingly statistical analysis of the quantitative data it was now possible to collect. Description had been replaced by investigation: natural history, it appeared, had now become science.

Hrdy had gone to the field to test the hypothesis that infant killing was a result of crowding, usually associated with human disturbance. Abu, with its large population living in close commensal contact with humans, was a good place to examine this. What she found there redirected her attention from the immediate context of infanticidal events to the ultimate consequences of that behavior for both the individual perpetrators and the maternal victims. Her initial reports of her experience described the context of her observations in great detail, incidentally providing her opponents with the ammunition they would use to undermine her case, but also placing them in an explanatory framework that provided a clear, logical, and comprehensive account of how and why such behavior might occur. In particular, she laid out a series of conditions under which males might be expected to commit infanticide and a series of questions about behaviors surrounding infanticide that merited further investigation (male discrimination of offspring, female counterstrategies, and so on). Her own observations were suggestive rather than conclusive and were eagerly challenged by those who found it difficult to recognize in their own animals the selfish strategists she described. Restricted from returning to India to take her own studies further, she had to leave to others the quest for more data against which to test the sexual selection hypothesis.

Fielding the Question, Questioning the Evidence

The years following the publication of Hrdy's work, and the initial skirmishes of the controversy, saw the steady production of more and more examples of behavior that appeared to fit the general context Hrdy had laid out for infanticide. In most cases these also included the raw observational (descriptive) data covering the infanticidal event, normally presented separately from the "discussion" sections recounting the writer's grounds for interpreting this act as adaptive. Individual attacks on infants represented only one part of this accumulating database: behaviors that were later treated as typical examples of potentially infanticidal situations were included, along with the experimental work being conducted on rodents and accounts of infant disappearances under conditions that met the predictions Hrdy had described. Very few of these examples could exactly match the "ideal type" of infanticide, but the gathering consensus seemed to be that the general level of "fit" between ideal type and reality was close enough to consider the matter closed: infanticide by male primates had evolved as a consequence of sexual selection.

The stringent criticisms of the hypothesis that were made during the 1970s and then again during the 1990s used similar strategies to discredit and destabilize this incipient consensus. First and principally, they returned to the detailed description of the contexts of the attacks to hammer it home that in almost none of the examples commonly cited to support the adaptive account of infanticide were the predictions and conditions of the sexual selection model met in full. On this reading, partial or incomplete "fits" did not and could not count and had to be excluded from the database (resulting in the very different tallies of the total number of observed infanticides). Second, they approached and defined the relevant literature in a completely different way. So, for example, while at the first eruption of the controversy the critique concentrated solely on the langur data, to the exclusion even of the other primates, at the second, Sussman and his colleagues did extend the critique to the primate data but excluded the rest of the comparative work.[17] Third, when dealing with evolutionary questions in the field, they adopted an approach that might be considered "horizontal" rather than "vertical."

Ultimately this was what the controversy was about: different conceptions of the role that evolutionary theory and evolutionary questions might play in understanding behavior, conceptions rooted in the changes in field practice that had taken place from the middle 1970s on and influenced by the implications for the study of human society. Supporters of the sexual selection hypothesis concentrated on the repercussions infanticide might have for those most directly involved—the male responsible and the infant-deprived mother—and considered the consequences that infant death might have for their lifetime reproductive success. This tactic took a vertical orientation to the study of evolution in the field, concentrating on projecting the results of infant killing into the future: if a male killed an infant, then in *most* circumstances he was likely to have invested in his own lifetime reproductive success, using the only currency the bank of evolution recognizes—offspring.[18] Opponents of the sexual selection hypothesis rejected this focus on the individual's future reproductive achievements and turned instead to the patterns of relatedness and the reproductive and mortality statistics recorded for the whole local population. This approach addressed the problem horizontally: attacks on infants had to be understood not as individual events but as part of a much larger picture whose details were still only imperfectly understood. In particular, knowing the nature and degree of relationships throughout a population would be the only way to ensure that males were no relation to the infants they attacked. Showing that these males had (probably) not fathered those particular infants answered only part of the question. Rather than looking to the future, opponents of the hypothesis argued that one should concentrate on the past and present of an infanticidal incident, questioning relationships and considering infant morbidity in general rather than focusing on a specific instance.

In line with this, by the 1990s these two perspectives could be linked with distinctly different approaches to evolutionary time. One of the major complaints by Bob Sussman, for example, was that supporters of the adaptive hypothesis went to the field expecting to see evolution in action. The concepts, hypotheses, and metaphors of sociobiology give the impression that every population is always in the process of adaptation: hence students were unconsciously primed to develop adaptive—ultimate—accounts of behavior even where they would be impossible to test, since "ultimate" explanations had more status that prosaic proximate ones. But if populations are adapting, they are changing: hence his demand that supporters of the hypothesis produce evidence of differential reproductive success and polymorphism.[19] If the population is adapting, then males who possess the "genes for" infanticide should indeed be leaving more offspring than males who do not. In this sense he is concentrating on the present and the future: the population is currently in flux, and the next generation(s) should contain more infanticidal males than the one being watched, otherwise adaptation cannot be occurring. In contrast, supporters of the hypothesis point back to the past. They reject the need for polymorphism or differential success, because they argue that the trait is fixed in those populations where it exists. All that is necessary for its perpetuation is that males avoid killing their own offspring and, as a result of infant killing, leave more offspring than they otherwise would have. The "traits for" or "genes for," or whatever biological basis the behavior has, developed far enough back in the evolutionary past for female counterstrategies themselves to have evolved in response—and as a result, infanticide, or the threat of infanticide, may still have a major influence over present-day behavior even where it has never been seen. In this sense the population is not "evolving"; rather, current behavioral strategies have been shaped by past selective pressures.

To sum up, despite the strangeness of the behavior and its positioning at the center of a cluster of topics that were of great interest not just to primatologists but to wider academic and lay publics, the topic of infanticide initially failed to be controversial. Although it was the subject of much discussion and many plausible explanations were offered, it coincided with then-current expectations about the relation between behavior and environment. Infanticide was likely to occur where social change was frequent and violent and was caused by intense population pressure or human disturbance, an unfortunate consequence of chaotic situations where unfamiliar males and females were fighting and mating in quick succession. However, the subject quickly became controversial with the emergence of a powerful new theoretical perspective that—regardless of its undoubted scientific success—did mesh closely with prevailing assumptions about the relation between men and women, sex and violence, dominance and success that were shared among Western cultures, and were expressed particularly strongly in the United States. This perspective also fundamentally changed the approach behavioral biologists took to studying animal behavior, both in the field and in the laboratory. There was one crucial reason infanticide could remain controversial for primatologists long after sociobiological theory (under various names) had largely triumphed elsewhere: the particular characteristics of field site research as it had developed within primatology over the middle to late twentieth century, or the fieldworker's regress, as it was expressed through the particularity of place.

Observing Nature, Witnessing Nature

For those who supported the sexual selection hypothesis, the paradox of the infanticide controversy was reflected in that the population density at certain field sites was indeed the underlying cause of the infanticidal behaviors also witnessed there. Carel van Schaik pointed out that "what we think is the causal factor, violent male turnovers, or at least replacement of dominant males, is much more common [at these sites]. So I always call it such a wonderful historical irony, that it was first described in this situation where you have much higher rates, artificially higher rates, but the phenomenon isn't artificial. And so they were both onto something." Dolhinow and others had rejected the use of sexual selection to explain primate infanticide in the first instance because crowding was, at that time, known to produce pathological behavior. It is indeed ironic that thirty years later their beliefs were borne out: but for supporters of the adaptive hypothesis, this is so *only* in that infanticide is more *likely* to be seen in disturbed populations that have been subjected to marked human influence. This *does not* mean it is the *product* of human influence; such stressed conditions only make it more likely that the underlying behavioral strategy of infanticide will be elicited. Observations at these disturbed sites gave rise to a hypothesis that enabled researchers not only to make sense of behavior there but also to develop further propositions about the wider evolution of male-female relationships and the selective factors that had driven the development of primate sociality. But these remain *hypotheses,* since the evolution of behavior cannot be observed in the field but can only be inferred.

Notably, the importance attributed to *seeing* infanticide changed dramatically with the controversy's development. Increasingly, primatologists took seriously Hrdy's (1977b) suggestion that the *absence* of observations of infanticide in certain species would need to be explained. Note that several attached great importance to having personally witnessed an infant being attacked and considered it possible that this made one more willing to take sexually selected infanticide seriously. For several researchers, having themselves seen an infant killed, whether in the field or in captivity, meant they were willing to give more weight to circumstantial evidence when assessing accounts of primate infanticide—*not* to conclude that sexually selected infanticide had *definitely* occurred, but to accept that *on the balance of probabilities,* it most likely had caused the infant's death or disappearance. Such judgments, apparently subjective though they might be, seem to correspond with Richard and Schulman's advice that sociobiological models, not amenable to testing in the same way the physical sciences are, "must be evaluated and judged more or less plausible on a variety of grounds" (1982, 248). Plausibility rather than demonstrability, making sense rather than showing outright, were and are central factors that make doing field science broadly different from doing science in the laboratory and distinguish the experimenter's regress from that of the fieldworker. As with the experimenter's regress, however, distinguishing between a plausible model and an implausible one depends on reaching some consensus on the relevant evidence by which the model must be evaluated.

Particularly in the early years of primatological research, and in relation to approximating and understanding the extent of primate variability, researchers debated—sometimes with a vehemence approaching that seen in the infanticide debate—the reasons behind observed differences in behavior between primates of the same species at different sites. Since the prevailing assumption was that animals should show species-typical behavior, then if a particular kind of behavior was seen at one site and not at another, must researchers at one have misinterpreted what they saw?[20] Charles Janson explained that these reactions had been common in the previous decades of primate fieldwork and suggested that initially

> there could have been some basis for it, in the sense that observations of course, were not very complete, they weren't reasonably well-habituated, long-term-studied animals, again because the life spans of these animals are so long. So it was always possible that people had in fact missed behaviors that might have been, quote unquote, chimpanzee universals, but I think eventually the mass of evidence suggested, well, no, this behavior is really absent in my population, it's not just that I missed it, it's really not there, and so demands an explanation.

In this sense one can see that seeking explanations for the absence of an event that most researchers saw as relatively well-documented made sense in the light of this broader shift.

Field sites are chaotic and variable places. In some environments observers can monitor the behavior of animals relatively continuously; in others it is much harder to keep track of what the members of a group are doing, or even to keep a count of how many animals are present at any given time. Learning to observe under these conditions involves, among other things, good self-management and personal endurance.[21] But even after mastering such skills—and doing so is a continuing process—one would have to be extraordinarily lucky to reliably complete scheduled observations as planned. And in the case of rare and fast-moving events, which necessarily must be observed ad libitum, recording everything that happens is not humanly possible.[22] But opponents of the sexual selection hypothesis were carefully following Carpenter's description of an "ideal type" field report in demanding this veridical reportage. In some cases, as in Hrdy's original reports of events at Abu, events were clearly being reconstructed in the absence of any *scientific* observers. But other reports were rejected because there was a lack of information pertaining to a particular aspect of the hypothesis, even where other predictions were being fulfilled. One way around this, and indeed, a strategy that primatologists increasingly adopted as the twentieth century progressed, was to recruit local field assistants who not only would provide extra eyes to witness as events unfolded but could remain at the field site relatively permanently, thus avoiding the problems created by breaks in observation such as Hrdy experienced at Abu. And if there are observers there, whether foreign scientists or local assistants, as Andreas Koenig argued, "If somebody has trained personnel, then everybody must believe this data. If you question these data, then I mean, what can you do?" Doing field science depends, finally, on trusting those who witness behavior to report it accurately.

Conclusion

At least for an outsider looking in, the infanticide debate is shot through with metaphors of seeing, invisibility, and blindness. Opponents of the sexual selection model castigate supporters for not being able to isolate the physical element that natural selection might be acting on to produce infanticide and for depending on invisible, or at least unseen, behaviors to justify using theoretical chains of logic to explain recorded events. Supporters of the model are infuriated by their opponents' demand that infanticide be subjected to the formal tests of adaptation that they feel cannot readily be satisfied in the field, and by their refusal to acknowledge experimental work as relevant even if it does meet such strict standards. But as I have tried to show, these divergent perspectives on the best way to account for animal behavior are not simply products of a particular controversy. They have their roots in the way behavior has historically been studied in the field.

Doing science in the field has from its very inception depended on the willingness of interested communities to assume that observers can produce reliable and relevant testimony and to take their word for its accuracy, As the introduction to this chapter suggested, the characteristics of such dependable accounts have inevitably shifted over the past thirty to forty years, partly through the elaboration and transformation of the theoretical frameworks for understanding behavior, but also because the field context itself has expanded, with concomitant changes in field practice. As the primatological literature accumulated, it became necessary to explain primate variety: researchers found they could not just describe behavior but had to explain why the behavior of "their" animals differed from the reports by other observers. This was not an easy position to be in, when observations could not be checked and description and interpretation were fully interpenetrated.

Given these peculiar but consistent characteristics of field science, it is perhaps surprising that the field is not constantly paralyzed by controversy. The explanation, as I have suggested, can be found in the fieldworker's regress: the constant awareness that conclusions are inevitably partial and that not everything can be accounted for. In this sense, the criticisms that—for experimental science—are applied to the work of opponents are key components of field scientists' working life and part of their tacit knowledge of what can and cannot be accomplished in this uncertain environment. Any criticism that could undermine another researcher's work could be applied to one's own results: opening the field site "black box" could be disastrous. Initially, the question of what was at the root of infanticide, or infant killing, was in exactly this position: several hypotheses were proposed, and while no definite conclusions were reached, none of the suggestions was truly controversial. It was the introduction of sociobiology, together with—*and just as important*—the period of reorientation in field practice that the sociobiological paradigm paralleled and provoked that lit a fire under this particular black box. The debates surrounding infanticide and infant killing were inevitably tarred by the wider sociobiological controversy, but this could not have become a scientific controversy without the particular characteristics of field science: the ongoing need to manage the field site as a scientific space and the shifts in the standards of doing field research that are unavoidable in an observational science.

CONCLUSION

There is no end in sight for the infanticide controversy. Although both opponents and supporters of the sexual selection hypothesis are willing to consider that they might be wrong, this has not been, and almost certainly will not be, translated into a rapprochement. The controversy itself is now so thoroughly interwoven with wider debates concerning the theoretical understanding of the evolution of primate sociality, the role of evolution in social life, and the implications these positions have for the conduct of field science itself that it is almost impossible to envisage how the different sides could talk on common ground—at least about infanticide. It is unresolvable in the classic sense of controversy studies: it is not that one side is wrong and the other right, or even—as participants suggested—that one side prioritized patterns while the other focused on detail. Instead, when participants' positions on the study of infanticide were placed in the context of their wider opinions on how to conduct field science, two broad—and fundamentally different—perspectives emerged. Where one side concentrated on the experiences of individual primates in the light of their recorded history and (possible) future, the other focused on the far wider problem of how local primate groups have interacted within themselves, with other primate groups, and with the local environment. I cannot and would not consider one position more accurate than the other, but I do suggest that it is this basic disjuncture—which resulted from *both* the introduction of sociobiological theory *and* the development of long-term field sites—that explains the intractability of the infanticide dispute.

For many, however, this may seem a very minor affair. Not only is it largely confined to the primatological literature, but it hardly consumes the working lives of even those most thoroughly caught up in it. The participants—at least in the later stages of the controversy—continue to work and publish in many other areas of primate life as well as in other arenas of behavioral biology. Are the infanticide debates then a minor footnote to the history of primatology? No. As this book has shown, not only do the debates reflect and focus far broader themes present within the discipline since its inception, but they also provide an opportunity to consider the conduct of field science in depth. In the first instance, it is evident that the conduct and experience of field science are very different from the conditions that pertain to laboratory work: in the second, it shows that we desperately need to know more about how field science works.

This specific controversy arose as different approaches to conducting science in the field developed, approaches based on a range of pragmatic and methodological, as well as theoretical, solutions to the problems inherent in working in this unenclosed and unconstrained environment. Although the question of place has for some time been central to the investigations of the history and sociology of science, and despite a steadily increasing interest in "science in the field" over the past decade or so, the laboratory still stands as the iconic site of scientific investigations. But over the course of the twentieth century it became clear that we need both laboratory study and field enquiry if we are to come to terms with some of the most pressing issues facing human societies. The history of primatology reflects this realization, as pioneering researchers like Yerkes and Zuckerman became conscious that if one wished to understand primate psychological and sociological development in full, one needed to complement laboratory attempts to study dyadic interactions by examining complex social systems in their environment of adaptation. The early years of primatological field research were characterized by intense debates between practitioners as to the best way of obtaining accurate and representative information where conditions were diametrically opposed to those traditionally thought of as scientific. The "ideal type" of the laboratory was the primary site of comparison, and early researchers were uncomfortably aware of the authoritative shortcomings of the field.

The partial solutions found to the problems of the field, together with the gradual emergence of the field site as itself a form of "scientific instrument," did much to contribute to the emergence of the infanticide debate. The tendency to treat one's own field site as the paradigm of that particular species' behavior persisted alongside the developing awareness of immense primate social variability; the continuing commitment to studying "natural" behavior was maintained even where researchers found it impossible to define a "natural" primate group; and the tendency to marginalize the dynamic development of the site's social and ecological conditions was found alongside the dawning awareness that only long-term research could provide the resilient answers researchers were pursuing. All these examples helped create the conditions under which infanticide and its evolutionary status were to become so problematic. The emergence of sociobiology and the concomitant hostility between the natural and the social sciences certainly made up an essential part of the controversy, but they were by no means the sole factors. In the 1960s, field methodology and dominant social theory kept infanticide noncontroversial: in the 1970s, theoretical controversy propelled it to the fore—but it could become and remain controversial only where there was fundamental fluidity in what constituted best methodological practice. By the 1990s methodological standards were again at the heart of attacks on the sexual selection hypothesis, but underlying this were at least two distinct viewpoints on how to study evolution in the field, which I have called the vertical and horizontal approaches, exemplified in the standard bearers of the controversy but existing to some degree throughout primatology.

Collecting information comparable to that produced within the laboratory from places that were so different was extremely difficult, and the notion of witness was central. As proposed in the work of Clarence Ray Carpenter, the question must be asked, "Who made the observations, and what were their qualifications for making them?" Later in the controversy this did not just refer to the fieldworkers' honesty or reliability but applied to their experience—and crucially—to their early training and how this framed their vision. From the inception of the discipline, primatologists had been drawn from all over the academy—from anthropology to zoology, comparative psychology to mathematics, sociology to anatomy—and many of the early debates were provoked by the need for one discipline to make sense of another's language and background. In such an atmosphere of interdisciplinary encounter, goodwill is necessary if the effort at comprehension is to succeed—and while it still needs to be oriented toward rigorous standards of testing, knowing what might or might not be possible within particular physical constraints can leaven a strict interpretation of what constitutes best practice. Commonsense understanding of what constrains field practice normally meant key "black boxes" remained unopened: most important, authors' claims that their field sites were places where natural behavior could be recorded. But the inevitable lack of agreed-on distinctions meant that the *possibility* of opening this question remained ever present.

One consequence was that primatologists were forced to become reflexive in accounting for their own practices. What was extremely noticeable when interviewing participants in the debate was their willingness and capacity to hold their own work at arm's length: to consider over and over the grounds on which they made their observations, the probable limitations to their vision (both literally and metaphorically), and the range of potentially contradictory interpretations that could be made. I would like to think it was my interviewing skills that produced such thoughtful and critical appraisals: unfortunately for my self-esteem, it is far more likely that it was the habitual uncertainty that attends primatological field practice, characterized in this book as the "fieldworker's regress"—perhaps intensified because the subjects of the primatological gaze are primates, with their close genetic and morphological resemblance to humanity.

To return to the earlier question, Does this mean that the conclusions drawn from the infanticide debate must be held as specific to primatology, or to primatological field science? In relation to the direct conclusions about the source of the controversy within the history of primatology and its interdisciplinary origins, yes. But at the same time, as I noted above, many of the most urgent problems now facing global society share formal characteristics with the infanticide debates. The solutions to global climate change, loss of biodiversity, and the genetic modification of organisms—to name just a few issues—will be found not just in the laboratory and in the field *but also* in the complex interplay between these two environments, as nature becomes artifact when it is transformed into something that can be understood within the laboratory walls, and as these artifacts are in turn translated into their potential consequences as they are introduced to the unrestricted global environment. But these solutions cannot be determined in isolation from their wider political, economic, and social context. The characteristics of their public circulation through their representation in the media and in political and administrative debate will be as important to their success or failure as will their scientific content.

In this context, sociologists and historians of science have a crucial role. As academics, we have largely refrained from playing the "public intellectual"—with some exceptions, of course.[1] The discipline's early commitment to methodological relativism and the principles of symmetry and impartiality—that is, to the idea that to rigorously investigate the problems of scientific practice one must treat both sides of a scientific debate equally—meant that we have largely avoided public intervention, though our work has sometimes been used in public debate. We have done our best to refrain from judgments on the truth or falsity of competing claims in a scientific debate and have instead concentrated on investigating the status of these claims as they move from one context and one community to the next.

Where the consequences of unrestricted human intervention in global ecosystems have become so serious and where scientific authority represents the ultimate source of legitimacy in public debate, it becomes disingenuous to avoid the consequences of our work in this way.

In emphasizing this point, I do not for a moment suggest that we abandon our tradition of impartiality. Instead, I urge far greater efforts to make the tools and the results of our work available and comprehensible to the general public. If one thing unites most of the programs for the history and sociology of science, it is the acknowledgment that "science" as a subject has been unjustifiably reified, that by and large people have either forgotten or chosen to disregard the fact that science *is* what scientists *do:* that is, it is the result of humans interacting both with each other and with the "natural" world. That science is a human undertaking makes it likely that subjective—that is, *social*—elements will be drawn on when making decisions, not just about the financial support of different projects, but also about the truth or falsity of scientific claims. As the study of scientific controversy has repeatedly shown, many scientists do acknowledge this tendency—but critically, only with respect to claims they consider erroneous. A colleague may well believe a false claim because he or she was trained in a particular laboratory, but accepting the truth requires no such explanation. When sociologists and historians pointed out that equally subjective reasons might lie behind belief in a "true" claim, it was unfortunately presumed that we were suggesting either that all science was false or that all truth claims were equally valid. We were not. And I hope with all my heart that this book has demonstrated that different scientific communities—even within a single discipline—establish their own standards for making judgments about relative validity.

The irony, however, is that the main accusation lodged against science studies programs during the science wars was that we were deliberately trying to destabilize the authority of science by drawing attention to the subjective context of its creation. But the point is that absolute objectivity is unattainable: scientists do disagree, and they base their disagreement on a whole range of grounds—as chapter 6 showed for the infanticide controversy. But it is exactly this subjectivity, combined with the general *expectation*—from both laypeople and experts—that objective decision making is possible, that provokes disgruntlement and delay when society faces a science-based problem that requires unpopular action. For example, politicians and policy makers have postponed decisions about the nature of the global threat posed by climate change by pointing to what appears to be a lack of scientific unanimity about the consequences of human intervention in natural environments. This lack of agreement apparently justifies waiting for a consensus on appropriate evidence.

However, social studies of science have shown that such unanimity—especially concerning an immediate or novel problem—is a pipe dream. Certainty is impossible to achieve—one can only depend on best estimates. Demanding it puts scientists in an impossible position. Whether we are considering the position of the person in the street, of civil servants, or of elected politicians, what people need to understand is how science really works, not the traditional hypothetical-deductive method we were taught in school. Speaking about science and treating scientists as if it and they had direct access to "the truth" does a disservice both to them and to the public, since it sets up expectations that cannot be met in the short or the medium term. Those of us in the science studies community need to do a far better job of explaining this and of working more directly with scientists to enable us both to understand them better and to be able to explain their actions to the wider community. What are anthropologists, sociologists, and historians for, if not to help different groups understand each other?

APPENDIX

Infanticide Interviews, 2002–3

Charlie Nunn	University of California, Davis	4/4/02
Sarah Hrdy	University of California, Davis	4/5/02
Kelly Stewart	University of California, Davis	4/5/02
Sandy Harcourt	University of California, Davis	4/6/02
Tim White	University of California, Berkeley	4/8/02
Bob Sussman	Washington University, St. Louis	4/9/02
Jim Cheverud	Washington Medical School, St. Louis	4/11/02
Craig Packer	University of Minnesota	4/15/02
Anne Pusey	University of Minnesota	4/16/02
Carola Borries	SUNY Stony Brook	4/19/02
Andreas Koenig	SUNY Stony Brook	4/19/02
Charlie Janson	SUNY Stony Brook (now at the University of Montana, Missoula)	4/21/02
Carel van Schaik	Duke University, Durham, NC (now at the University of Zurich)	4/22/02
Leslie Digby	Duke University, Durham, NC	4/23/02
Phyllis Lee	Cambridge University, UK (now at the University of Stirling, UK)	11/3/03
Tim Clutton-Brock	Cambridge University, UK	11/3/03
Thelma Rowell	University of California, Berkeley (retired)	11/6/03
Volker Sommer	University College London, UK	11/13/03
Louise Barrett	Liverpool University, UK (now at the University of Lethbridge, Alberta)	11/23/03
Robin Dunbar	Liverpool University, UK (now at the University of Oxford, UK)	11/23/03

NOTES

INTRODUCTION

1. A note on terminology may be appropriate here. Opponents of the sexual selection hypothesis have preferred to use "infant killing" rather than "infanticide," arguing that the latter term begs the question at the heart of the controversy.
2. For details see the "trilogy" of edited infanticide volumes (Hausfater and Hrdy 1984; Parmigiani and Vom Saal 1994; Van Schaik and Janson 2000). At present, adults have been reported to attack infants in amphibian, fish, bird, rodent, and carnivore species as well as primates: however, while most of the explanations put forward for these intraspecific attacks are adaptive rather than pathological, note that the sexual selection hypothesis is not the only example of an adaptive account of infant killing. The range of adaptive explanations were specified in Hrdy (1979) and will be further discussed in chapter 5.
3. At the turn of the millennium, largely unsuccessful attempts were made to widen the controversy to include the studies of lion social behavior. I will consider these briefly later in this chapter and in more detail in chapter 5.
4. The "science wars" began in the mid-1990s when a number of academics and authors began to protest what they saw as the misrepresentation of science by historians and sociologists. Prominent in these attacks were Paul Gross and Norman Levitt (1994), who in 1996, with a number of other academics, organized a conference sponsored by the New York Academy of Sciences, titled "The Flight from Science and Reason" (Gross, Levitt, and Lewis 1997). In the same year, possibly the most famous event of the science wars took place—the publication of Alan Sokal's hoax paper in the journal *Social Text* (1996a, 1996b)—and with Jean Bricmont, Sokal went on to develop his attack on what they characterized as "postmodernist" critiques of science (Sokal and Bricmont 2003/1998). In response, those involved in science studies defended their subject vehemently (for example, Ross 1996; Fuller 1997). For a time it appeared that continued polemic was the only prospect, but the publication of Labinger and Collins's edited collection of papers (2001) demonstrated that it was at least possible for scientists and nonscientists to have a conversation about this topic.
5. A full bibliography of these investigations would take far too much space. For examples of examinations of ongoing controversies, see Collins and Pinch's work on parapsychology (1982), Mulkay (1997) on embryo research, Martin on fluorida-

tion (1991), and Collins's updates of his original work on the gravitational waves dispute (1999, 2004), among many, many others. Some examples, such as the cold fusion saga, occupy a slightly less clear-cut position: Lewenstein (1995) and Gieryn (1992) both treat the controversy as resolved, but Simon (1999, 2002) indicates that there may yet be life in it. Examples of investigations of historical controversies are equally numerous. See, for example, Rudwick 1985; Shapin and Schaffer 1985; Latour 1988; Cole 1996; Collins 1992/1985; and MacKenzie 1990.

6. As a theoretical problem, this question has been most usefully considered by Jasanoff (1996) and, more recently and more polemically, by Fuller (2006). However, many other authors have described their disquiet at the use that has been made, or may well be made, of their work (Collins and Pinch 1982; Scott, Richards, and Martin 1990; Richards 1996).
7. Currently there is an immense literature of Kuhnian scholarship, and interpretations of both his arguments and his legacy vary widely. See, for example, Barnes 1982, 1990; Bird 2000; Cohen 1985; Fuller 2000; and Nickles 2002.
8. The "strong program" was so named in contrast to previous "weak" programs for the sociology of science. Heavily influenced by Wittgenstein and his approach to the sociology of rules, the community of philosophers, sociologists, and historians at Edinburgh in this period began trying to use the *context* of scientific knowledge to understand its *content*. Rather than seeking only external explanations of scientific activity, for the first time the internal workings of science as a process, as an institution, and—eventually—as a system of interactions came under detailed scrutiny (Bloor 1991/1976; Barnes and Bloor 1982; MacKenzie 1981; Shapin 1975; Barnes, Bloor, and Henry, 1996).
9. Collins's work, of course, also drew on ethnographic research (Collins 1992/1985; Collins and Pinch 1982), but for examples of detailed ethnographic work see Knorr Cetina 1983; Lynch 1985; and Traweek 1988.
10. That is to say, knowledge that cannot readily be communicated verbally but is spread through demonstration rather than explanation. See Collins (1992/1985).
11. At the simplest level, for example, researchers working in different laboratories will have access to slightly different equipment.
12. See, for example, Franklin 1990.
13. For examples of how the emergence of the laboratory as a social and physical space functioned as a key development in the creation of modern modes of scientific inquiry, see Hannaway 1986; Ophir and Shapin 1991; Secord 1994; and Shapin 1988, 1994.
14. One of the first, and certainly one of the most detailed, accounts of a controversy in field science was provided by Martin Rudwick (1985). More recent examples of studies of field science in general include Crist 1996; Kuklick and Kohler 1996; Latour 1999; Roth and Bowen 1999; Henke 2000; Rees 2001b; Kohler 2002a, 2002b; and Gieryn 2006.
15. I will explore these issues in detail in part 1 and return to them frequently throughout the text.
16. I am indebted to an anonymous reviewer for the term fieldworker's regress.
17. It should, of course, be noted that Wilson did not himself develop the key concepts that underpinned his sociobiological synthesis: that work was done by William Hamilton, Robert Trivers, George C. Williams, and John Maynard Smith and will be discussed in more detail in chapter 4. He was, however, responsible for disseminating them to a far wider audience, not least because of his decision to apply sociobiological concepts to human social structures.

18. See the appendix for a list of all interviewees and the dates and locations of interviews. In total, I interviewed twenty-one primatologists, most of whom (fifteen) were either based in North America or had spent a significant proportion of their academic lives working there. Initially I approached people who were directly involved in the infanticide controversy; inevitably some either declined to respond or, having responded, were unwilling or unable to participate in the project. As I visited universities, other researchers made themselves available to discuss the controversy with me. With the agreement of the participants, each interview was taped on the understanding that no material would be used without the approval of the interviewee. I transcribed the completed interviews and sent copies to the inteviewees for correction. One interview had to be discarded at this stage owing to the poor quality of the recording. Finally, all quotations from the interview transcripts have been seen and approved by the individual speakers. When no other source is given for a quotation used in the text, that material is drawn from one of the interview transcripts. For a more general discussion of the methodological issues involved in the interview process, see Rees 2001a.
19. For example, had sociobiology changed the way researchers thought about studying primate behavior, or just the aspects of primate behavior they chose to study? Was it possible to operationalize sociobiological concepts adequately? How do you measure fitness in the field? How much and what kind of evidence do you need to justify applying a sociobiological explanation? In particular, how could the applicability of sociobiological theories to primates be tested, given the animals' long life and the difficulty of studying them in the wild?
20. The publication of *Sociobiology* in the summer of 1975 was followed in the autumn by a letter to the *New York Review of Books* (Allen et al. 1975) accusing Wilson of biological determinism and of resurrecting ideas linked with the eugenics programs that had existed in Europe and North America in the first half of the twentieth century. Wilson responded by asserting that he had deliberately been quoted out of context (1975b). The debate soon spiraled; see Segerstrale (2000) for details.
21. This approach conflicted sharply with what had previously been known and believed about animal behavior. For example, the work of Konrad Lorenz, one of ethology's "founding fathers," had specifically excluded the idea that animals could, in normal circumstances, show lethal violence toward each other (Lorenz 1963), while John Calhoun's (1962a, 1962b) experimental production of the "behavioral sink" in laboratory rats had demonstrated that artificially crowding animals could produce very high levels of aggression and infant killing. Both these influential models assumed levels of intra-animal aggression that could be fatal were by definition abnormal.
22. Sociologists and historians of science have demonstrated that the popularizing, or rather publicizing, of science, far from being the transmission of complex ideas in a simplified form from the expert to the layperson, represents a key stage in the making of scientific knowledge (Fleck 1979/1935; Rudwick 1985; Shapin 1988; Collins and Pinch 1993; Golinski 1998). In addition, Nelkin (1985), Myers (1990), and Lewenstein (1995) also demonstrate that popularizing is a key strategy used by scientists proposing a new and controversial approach to a particular field.
23. The official tally of infanticides is an index of how far the study of this phenomenon was embrangled in intense dissension—at this stage, Hrdy argues that thirty-nine infants have disappeared in the context of male takeovers (that they have most probably been killed by usurping males) and that this is a *conservative* estimate

(1977b, 46), in contrast to the three that Curtin and Dolhinow are willing to acknowledge. Fifteen years later, when the controversy erupted for the second time, while for supporters of the adaptive hypothesis the number of known infanticides lies in the upper hundreds, Bartlett, Sussman, and Cheverud (1993) are willing to accept only forty-eight of these as actual examples of infant killing.

24. As later chapters will show, the case of the chimpanzees was to prove a difficult problem for supporters of the sexual selection hypothesis.
25. Detailed examination of these will have to wait until chapter 4. Examples of publications broadly supportive of the adaptive approach include Angst and Thommen 1977; Goodall 1977; Roonwal and Mohnot 1977; Struhsaker 1977; Chapman and Hausfater 1979; Packer 1980; Busse and Hamilton 1981; Butynski 1982; Sekulic 1983; Clarke 1983; and Winkler, Loch, and Vogel 1984. More critical accounts include Makwana 1979; Mohnot, Gadgil, and Makwana 1981; Boggess 1979; and Schubert 1982. Alternative, and in some cases potentially complementary, explanations for adult aggression toward infants include Rudran 1979; Ripley 1980; Kawanaka 1981; Rijksen 1981 Hausfater and Vogel 1982; and Ciani 1984. I will be examining these accounts in detail in later chapters, but note that they deal specifically with the primatological literature, not with work on other animal species.
26. Volker Sommer (2000) describes Vogel's conversion to the adaptationist approach, and see the discussion in chapter 5 for the particular circumstances surrounding the study of Jodhpur's langurs.
27. Run privately, and based in the United States, the Wenner-Gren Foundation was established and exists to support basic research in all areas of anthropology, including primatology and the study of human origins.
28. In the edited collection of conference proceedings (Hausfater and Hrdy 1984), the contents pages divide the papers into four sections: taxonomic/background reviews, primates, rodents, and humans. But *only* in relation to the primates (nonhuman) were problems with the adaptive hypothesis signaled—by use of a subtitle, "A Topic of Continuing Debate" (Hausfater and Hrdy 1984, vii). While even human infanticide was treated as having clear adaptive causes (Webster 1982), evolved infanticide in nonhuman primates was still disputed, although only one of the eight substantive papers in this section directly attacked the sexual selection hypothesis.
29. For a further example, compare Hrdy (1979) with Sherman (1981). Both are reviews of the then current literature on adaptive infanticide, but Hrdy, dealing more with primate species, is forced to spend a considerable proportion of the article addressing the controversy surrounding sexually selected infanticide in primates. In contrast, Sherman's work, which focuses on animals in general, mentions the controversy only twice—once in the introduction, and once to cite the work of Curtin, Dolhinow, and Boggess as an alternative view that links infanticide with habitat disturbance. Their grounds for objecting to the adaptive hypothesis are not given at all.
30. Again, selected papers will be examined in detail in later chapters, but examples here include Busse 1985; Nishida and Kawanaka 1985; Smuts 1985; Sommer and Mohnot 1985; Struhsaker and Leland 1985; Takahata 1985; Newton 1986; Agoramoorthy 1986; Sommer 1987; Newton 1987; Tarara 1987; Newton 1988; Hiraiwa-Hasegawa 1987; Dunbar 1988; Agoramoorthy and Mohnot 1988; Agoramoorthy et al. 1988; Gomendio and Colmenares 1989; Vogel 1988; Watts 1989; Breden and Hausfater 1990; Butynski 1990; Van Schaik and Dunbar 1990; Valderrama, Srikosamatara, and J. G. Robinson 1990; Yamamura et al. 1990; Pereira and Weiss 1991;

Boer and Sommer 1992; Hamai et al. 1992; and Smuts and Smuts 1993. Again, this is not intended to be an exhaustive list of reported infanticides.

31. Indeed, most of these infanticides had been seen at one site, Jodhpur, hardly surprising since only at that site had long-term observations of this species been continued.
32. For rodents, see Labov et al. 1985 for a review: for carnivores see Hoogland 1985 (prairie dogs); Pusey and Packer 1987; Hanby and Bygott 1987; Packer et al., 1988; Packer, Scheel, and Pusey 1990; and McComb et al. 1993 (lions): for birds, see Freed 1987 and Veiga 1993. Würsig proposes (1989), in relation to the aggressive attitude adult male dolphins displayed toward young, that "infanticide is common in a number of species and could occur in dolphin species in nature" (1154). Again, this is not intended to represent a complete list.
33. Hausfater (1984) suggested this was the mark of primatology's disciplinary maturity and intellectual cohesion: despite their disparate disciplinary heritages, the vast majority of primatologists, if not all, could now agree on a sociobiological disciplinary paradigm.
34. As chapters 1 and 2 will show, since the inception of primate studies in the 1930s, researchers have striven to explain why primates, unlike most other mammals, form relatively permanent bisexual groups.
35. Relevant here are the points made earlier in relation to the impact sociobiology had on primatology: consistently, the supporters of the sexual selection hypothesis I spoke to identified anthropology as a discipline that took an overwhelmingly descriptive approach to data collection, avoiding generalizations and systematics in favor of retaining the individuality of a given case study. This should not, however, be taken to imply that primatology in the United States was any more "descriptive" than it was elsewhere.
36. Again, for researchers in the United States, one result of the postmodern turn for anthropology had been a rejection of science itself, even of the word "science" to describe any part of their work.
37. I will explore this argument in detail in chapter 7, but for now it is important to make a distinction between "biological" thought, "evolutionary theory," and sociobiology/behavioral ecology. None of those who opposed the sexual selection hypothesis did so on the grounds that they opposed evolution: quite the contrary, several argued that those who held to the hypothesis had fundamentally misunderstood Darwinian evolution.
38. Van Schaik and Janson were to edit the third collection of papers on infanticide that was published by Cambridge University Press in 2000.
39. The first reference cited infanticide simply as an example of the application of sociobiology to primatology and noted that it was "controversial" (Strum and Fedigan 2000, 21). The second was in a paper contributed by Bob Sussman, who again summarized the 1993 paper to demonstrate the apparent lack of primatological evidence for infanticide and compared the case to the hoax of Piltdown man—infanticide, he argued, is based "more upon preconceptions about nonhuman and human behaviour than upon existing data" (Sussman 2000, 98). Finally, Van Hooff's chapter on the development of primatology in the Netherlands noted that the "importance of infanticide is still under dispute" (2000, 135).
40. Interestingly, he had been appointed editor in an attempt to "heal a growing rift within the discipline" of anthropology, a rift ascribed to the postmodernist turn (Shea 1999, 25).

41. Haraway 1990; Strum and Fedigan 2000.
42. This was not least the result of the influence of Sherwood Washburn, whose contribution to United States primatological history will be further discussed in chapter 2.
43. The women's movement was mentioned several times here, as was the impact on anthropology of postmodernism and what was considered its commitment to denying the existence of "fact."
44. That is to say, the attempt to draw ethical conclusions from one's observations of the natural world: the fallacy that what is "natural" is therefore also "ethical."
45. Some confined the usefulness of the critique to the very early stages, when it encouraged participants to distinguish between observed and inferred accounts of infant killing; others suggested that criticism was indispensable in maintaining high scientific standards.

CHAPTER ONE

1. Robert Yerkes was central to the development of the field of comparative psychology in the United States, and the study of ape and monkey psychology was, in turn, fundamental to his contribution to that development (Thomas 2006). He was responsible for creating what eventually became the Yerkes Primate Research Laboratory in Orange Park, Florida, where Solly Zuckerman, an anatomist from South Africa, was briefly employed before returning to academic life in the United Kingdom. Zuckerman went on to have a much wider influence over the general science policy of the United Kingdom, eventually becoming chief scientific adviser to the government (Burt 2006; Peyton 2001).
2. See, for example, Kohler (1925) and Yerkes (1943). Reynolds (1967) provides an overview of laboratory research in the early and middle twentieth century.
3. Zuckerman, for example, singles out Herbert Spencer as a particular sinner in this regard, using him to demonstrate his general point that "sociologists have found it easy to convince themselves that the behaviour of apes supports whatever view they may happen to hold about the origins of human society" (1981/1932, 13).
4. Not that this was confined to the early part of the twentieth century, of course. Landau (1984, 1991) and Latour and Strum (1986) have illustrated the role that subjectivity and selection can play in constructing narratives of human social origins.
5. Zuckerman famously concentrated on the behavior of the captive Monkey Hill baboons at London Zoo, but even he paid a visit to the field to see if his conclusions about the basis of primate social life were borne out. The theme of the need to examine "natural" behavior in primates or other animals in order to understand the basis of human social origins echoes through the literature of the second wave of primate studies in the twentieth century—see, for example, Scott 1950; Nissen 1951; Washburn and DeVore 1962; Jay 1962; Kortlandt 1960; and Washburn and Hamburg 1965.
6. Over this period, a number of edited collections appeared (Southwick 1963; DeVore 1965; Jay 1968; Dolhinow 1972), each intended to stand as a summary of best practice in primatology at that moment. Other publications tended to focus on particular aspects of primate lives: for example, Rheingold (1963) on maternal behavior, Napier and Napier on systematics and classification (1970), and Holloway on aggression (1974). This chapter will focus on the former accounts.
7. For example, Hall pointed out that while concepts such as dominance and territoriality had specific and technical meanings in relation to other areas of zoology,

during the early period of primate field studies they could be used only descriptively (Hall 1968a, 11). To get around this problem, people were encouraged to give precise descriptions of the physical actions they were observing, rather than simply naming a category of behavior, so readers could decide whether this resembled activities at their own sites (Jay 1968, 177). See also Crook (1970, 104–6) for an attempt to clarify the range of terms used to refer to primate groupings.

8. There had been at least one previous sustained attempt to study chimpanzees in their natural environment—the linguistic studies of Richard Garner at the turn of the twentieth century. See Radick (2005) for an account of this work and its subsequent marginalization.
9. Carpenter's conclusions about the nature and basis of primate sociability will be dealt with in the next chapter: here I will concentrate on the strategies he used to draw these conclusions.
10. Because of the significant influence that Carpenter's studies had on the development of primatology later in the twentieth century, the papers he wrote from 1934 to 1958 were reissued by Pennsylvania State University Press in 1964. References are to the new volume. See note 19 for a more detailed description of this volume's origins.
11. During the same period Carpenter had also watched spider monkeys in Panama, and he went on with varying degrees of success to observe orangutans in Sumatra and gibbons in Thailand (then Siam) and to collect rhesus monkeys in India to establish a monkey colony on Cayo Santiago (an island off the coast of Puerto Rico). The establishment of breeding colonies, at which research might also be carried out, represented another stage in Yerkes' plans for the development of primatology (Yerkes and Yerkes 1929, 587).
12. See, for example, Chance and Mead (1953) and Chance (1955, 1956). This work pursued Zuckerman's interest in the mechanisms and functions that lay behind the continuous association between the two sexes found in primates and will be discussed further in chapter 2.
13. A number of factors may have been responsible for this relatively sudden resurgence of interest in the lives of wild primates. United States affluence in the postwar years played a part, as did increasing access to cheaper and faster international travel; and it was during this period that key figures such as Washburn and Imanishi began to propound fundamental concepts such as the "new physical anthropology" and the "species society" that were to have a decisive influence on the way primatology developed in Japan and the United States (Washburn 1951; Imanishi, 1963/1957). In addition, in these years European researchers could take advantage of the colonial networks of contact and control that still existed throughout much of Africa and Asia (despite formal independence). Many of the early accounts of primatological research in the field written and produced for a popular rather than a professional audience reveal the extent to which researchers depended on such colonial social networks (see, for example, Schaller 1965c; Reynolds 1965; Goodall 1971; Fossey 1983).
14. An unattributed communication in *Folia Primatologica* (Anon. 1964) describes the founding of the Regional Primate Research Center program in the United States from 1956 (when the possibility of primate research centers was first discussed by the National Advisory Heart Council) to 1963. Congress approved funding for the centers in 1960–62, and seven were established, in Portland, Oregon; Seattle, Washington; Madison, Wisconsin; Orange Park, Florida; Covington, Louisiana; Marl-

boro, Massachusetts; and Davis, California. While each had its own investigative specialty, all were committed to using primates in research. A preliminary history of the primate research centers can be found in Dukelow (1995).

15. The journal switched to English after the first two issues but still remained largely unknown to Western researchers. In fact, articles in *Primates* continue to attract less attention and status than those in other journals (Asquith 2000; Takasaki 2000).
16. DeVore and Lee 1963; Reynolds 1963; Butler 1966a, 1966b, 1967; Davenport 1967; and Bernstein 1967.
17. The focus of primate research was still overwhelmingly on large-bodied terrestrial primates in Africa and Asia, however. For example, from 1970 to 1974 only seven out of nearly seventy articles on field research in *Folia Primatologica* and *Primates* deal with New World monkeys, and only three discuss prosimians.
18. Note that by 1974 a few sites in East Africa had in fact been research active for a number of years. For example, Goodall had begun work at Gombe in 1960 and the Kyoto University African Primatological Expedition had selected Mahale as a research base in 1965, but these sites were very much in the minority. The next year, however, after Zairian rebels kidnapped Western students from Gombe, the Tanzanian government barred foreign researchers from the site. Long-term work at Gombe was able to continue only because the Tanzanian field assistants maintained the research records in their absence (Goodall 1979, 1986).
19. For example, in 1964 Pennsylvania State University Press published a selection of his papers from 1934 until 1958, compiling monographs and articles that had become "part of the basic scientific literature of the field of primatology and of other related sciences" (Carpenter 1964, vii). This collection was thought necessary because of the difficulty of finding the original articles and their importance to the developing field.
20. Here it is useful to note the decline of collecting expeditions and the rise of surveys of primate populations over this period. Although both Zuckerman and Carpenter shot and captured animals to make more detailed examinations, this practice waned and was replaced by examining animals that had died either naturally or inadvertently (Dart 1960) and by increasing noninvasive surveys of primate populations (Frisch 1959; Tappen 1960, 1964; Southwick, Beg, and Siddiqi 1961a, 1961b, 1965; Simonds 1962). In a move that paralleled this development and also echoed the dependence on habituation as a key field research technique, researchers became increasingly concerned to ensure that "their" animals weren't hunted by the local people either.
21. This was published in a volume based on a nine-month "Primate Project" organized by Sherwood Washburn and David Hamburg, supported by the National Institutes of Health and held at the Center for Advanced Study in the Behavioral Sciences, Stanford, California. The project, and the resulting volume, was intended to "systematically survey the results of recent field studies" and present a unified account of the current state of primatological field research that would stand as an example of primatological best practice not just for students, but for other interested professionals.
22. These thoughts are echoed in Schaller's appendix to the same volume, which specifically outlines field procedures for primatology. Here he suggests that a field study would include, first, an ecological survey of the range; second, an account of the social life of a selected group; and third, intensive studies of a particular aspect of behavior. He also warns against the potentially subjective nature of qualitative judg-

ments. He proposes that the studies recommended in the third point might be supplemented by experimental procedures, either in the field or in the laboratory—but he acknowledges that those carried out so far have been overwhelmingly descriptive (1965b, 623).

23. This strategy corresponds with those identified by other researchers in the history of scientific expeditions and natural history, who have pointed to the way fieldworkers have made a virtue out of necessity, combining their limited use of laboratory technique with a celebration of place. See, for example, the accounts of Humboldtian science after Susan Faye Cannon's characterization of the influence of Alexander von Humboldt (Cannon 1978; Browne 1996; Dettlebach 1996), or Robert Kohler's more recent analyses of the conduct of field biology in the early to middle twentieth century, which stresses that field scientists used the uniqueness of each field site to bolster their claim to be making authoritative observations (Kohler 2002a, 2002b).
24. See also Lancaster and Lee 1965; Marler 1965; Hall 1968a, 1968b; and Jay 1968.
25. This is not surprising, in that the laboratory was, and possibly is, *the* premier site of scientific research. It is the place where modern science emerged, and other sites of knowledge production will understandably suffer by comparison: they must justify their lack of resemblance to the laboratory. Moreover, this situation is exacerbated for fieldworkers who carry out their research at some distance from the centers of Western knowledge production. The history of colonial science includes many examples of the way scientific authority clearly lay at the metropolitan centers, not on the peripheries. For example, see the Australian zoological controversy documented by Duncan (1987), which demonstrates that European researchers were unlikely to accept unusual empirical evidence unless it was vouched for by someone with a European reputation, or Browne's (1996) account of Joseph Hooker's attitude toward colonial workers (313). It is clear that the observations made and the materials collected by travelers to Africa, Asia, or the New World were historically of the greatest import to domestic natural philosophers (see, for example, Carey 1997; Findlen 1994: Shapin 1994). It is equally clear that by the late nineteenth century most travelers were personal and social unknowns who might well have had a vested interest in exaggerating their information to enhance its value. Since they were not themselves scientists or necessarily gentlemen, they were not directly subject to the sanctions of the scientific community, which therefore had no reason to trust them (McCook 1996). And yet, since their information was so important, some way had to be found of mediating between travelers and the scientific community. One way was to accept their observations but not their interpretations and to try to direct their observations from a distance through questionnaires to ensure that they were watching what the educated metropolitan elite wanted to see (Carey 1997). Empirical observation, while valuable, was meaningless unless placed in an intellectual framework, which travelers were assumed to lack.
26. Schneirla was writing in the *Annals of the New York Academy of Sciences,* in a volume inspired by the need for more systematic studies of naturalistic animal behavior and the desire to indicate some appropriate standards of investigation in this developing field.
27. See note 13 in the introduction.
28. Numerous writers have documented the way field science has borrowed or conscripted social and cultural practices from the people they share their sites with, as well as transplanting some from their own cultures of origin. Histories of geology and anthropology have both emphasized the flavors of heroism and romance,

even religion, to be found in the early accounts of fieldwork. Roy Porter discusses the "romance of the field" that developed among geologists in the Victorian period and suggests that "fieldwork became an explicitly religious experience, a spiritual re-creation" (1978, 820). Following Gupta and Ferguson's assertion that genuine fieldwork should involve hard work and physical danger, Kuklick argues that a key influence on fieldwork for anthropology could be found in the idea that "personal growth (of an implicitly masculine sort) could be effected through pilgrimages to unfamiliar places, where the European traveller endured physical discomfort and (genuine or imagined) danger" (Gupta and Ferguson 1997; Kuklick 1997, 48). Other writers have pointed to the effect that patterns of recreation, such as travel and sport, have had on the cultures of the field. Alpine scientists were also mountaineers (Hevly 1996; Robbins 1987); meteorological observations were made in hot air balloons that were also used for champagne picnics (Tucker 1996); and naturalists and collectors hunted and shot specimens for public display and private examination (Camerini 1996; McCook 1996; Haraway 1990).

29. In some later autobiographical writing by field primatologists one can chart clear shifts in identity from the observing scientist to the conservationist-advocate. See, for example, Fossey 1983; Campbell 2000; Goodall 1971, 1990; Goodall and Peterson 1993; and Strum 2001/1987, among others.
30. Behavioral biologists use this term to refer to animals belonging to the same species.
31. For a discussion of the problems of observing arboreal primates and the bias toward terrestrial animals in producing accounts of primate social structure, see Chalmers (1968a, 1968b) and Struhsaker (1969). The other reason for the focus on predominantly terrestrial primates was, of course, the feeling that these animals would provide a better model for understanding human biological and social evolution (see Washburn and Hamburg 1965 and DeVore 1963).
32. One of the reasons given for provisioning the animals was that it made it easier for tourists *as well as* scientists to watch them—and tourism represented a key source of income for the JMC in the early days. In addition, Walter Baumgartel, the hotelier who first drew scientific attention to the mountain gorillas in Uganda, tried to attract tourists to his hotel by distributing food that he hoped might entice the gorillas to visit too (Dart 1961).
33. This criticism was raised in relation to the Gombe chimpanzees; for example, see (Reynolds 1975). At least one writer (Ghigileri 1988) has suggested that Goodall's decision to provide bananas to encourage the chimps to remain close to her camp, and the later withdrawal of the feeding stations, might have been responsible for the violence that erupted between the Kasekela and Kahama chimpanzees.
34. Note that habituation is a two-way process that has continued to develop and change. During the later stages of primatological research, however, the danger was less that fear of humans would affect animal behavior than that the animals might seek to enlist the neutral human observers in their conflicts with other individuals or groups. See Rees (2007).
35. Having carried out fieldwork on primates and other animals for several decades, Tim Clutton-Brock made much the same point: "If an animal is worried about you, it's looking at you all the time, and you are quite likely to be affecting its behavior, particularly if it's starting to run away from you. And what you really want are animals that are entirely habituated to you, so that the further you push habituation, the less that becomes a problem, but what [strangers] actually see are very well ha-

bituated animals with [researchers] working very close to them, and they assume that it's much more likely that those people are affecting the animals' behavior. Actually, if you see people working a hundred metres away from the animals, they're much more likely to be affecting the animals' behavior. The closer they're working to them, the more the animals are accepting them and are unlikely to be bothered by them."

36. The "history of contact" varied tremendously, of course. In some cases animals had been hunted in the recent past (Ghiglieri 1988), in others they were provisioned by the local people, either voluntarily or involuntarily as a result of crop raiding (Strum 2001/1987). Some animals, such as baboons, were such ubiquitous raiders that the locals defined them as vermin to be shot on sight, while others, such as Hanuman langurs, were venerated. What was constant in these diverse human-primate relationships is how much care researchers working on such populations had to take in accounting for these relationships when reporting on behavior, usually in hopes of minimizing their potential effects—not so much on the animals as on the authority attached to their reports.
37. There are some similarities between this situation and the case of anthropology in the first part of the twentieth century. Like primatology later on, anthropology was undertaken to learn more about human social origins by studying the society and culture of groups of people thought to live a prehistoric (hunter-gatherer) lifestyle. The problem some anthropologists faced was that the very structures that made their research possible (the development of colonial networks of research, administration, and exploration) were destroying or mutilating the "simple" lifestyles they had come to study (see Stocking 1983, 1991; Kuklick 1991; Tomas 1991; and Grimshaw and Hart 1993 for examples). The persistence of this attitude into the postwar period has been illustrated by the work of Ruth Philips, who documented the exclusion of art objects made for tourists by the American Indians from museum displays; despite the age of some of these objects, they were considered contaminated artifacts because they had been made as part of the process of entering and negotiating European commercial and artistic systems (Philips 1994).
38. See Rees (2006a) for a discussion of the problems researchers faced in tracking animals in different social and ecological contexts.
39. Southwick and his colleagues found, for example, that it was impossible for researchers to maintain voice contact through the radio because it generated such intense interest from the local people: crowds would collect to watch what they did, preventing them from gathering data (Southwick, Beg, and Siddiqi 1965, 119). However, in the same study, the researchers used local informants to discover where the monkeys might be found, as Schaller did in his survey of the gorilla population of the Virungas (1965a, 325–27) before he was forced to leave Kabara in 1960 because of the revolution in the Congo. Information obtained from local inhabitants appears relatively frequent in this period, in relation either to discovering the location of nonhuman primate groups or to reporting rare and spectacular events such as deaths (for example, see Simonds 1965, 179).
40. This was not a novel development: another of the reasons for founding and maintaining the Cayo Santiago rhesus colony was to ensure a supply of macaques for use in United States biomedical research in case India restricted or barred their export.
41. Collias and Southwick (1952) attributed this decline to a yellow fever outbreak on the island, but later writers suggested it may have resulted from the unstable operation of a sexually selected infanticide strategy.

42. For example, they would offer a preferred food item to a pair of animals, and the animal that secured it would be considered dominant over the other. The appropriateness of this strategy was later questioned, as researchers such as Rowell (1974) pointed out that this definition of dominance was circular (access to a particular food item was considered to define dominance, and that the animal that succeeded in obtaining it was therefore dominant) and was also restricted.
43. Translocation experiments involved capturing a monkey from one group and transferring it to another. Kummer and his coworkers used this to examine which sex was responsible for maintaining the close physical relationship between the male and his harem of females. Later, in his autobiographical account of his researches, Kummer regretted being responsible for such manipulations (Kummer 1995).
44. The difficulties and dangers of manipulating "natural" populations are discussed in some detail in chapter 6.
45. Solly Zuckerman, for example, did not entirely approve of what he considered a premature move to quantification. In the second edition of *The Social Lives of Monkeys and Apes,* he suggested that some of the things people choose to count are meaningless, arguing that "some field workers who have been lured to the jungle appear to resort to figures which do little more than either confirm the obvious or add to the inventory of trivial observation. The statistical analysis of inexact measures may add a spurious air of exactitude, but it also soon evokes the suspicion that one is being 'blinded' by science" (1981/1932, 371).
46. Aldrich-Blake (1970) discusses this problem with regard to the observation of forest monkeys, as does Chalmers (1968a, 1968b). A much later example—the chimpanzees of the Tai Forest (Boesch and Boesch-Achermann 2000)—may illustrate this issue best: these chimps are perhaps most famous for using stone "hammers" as nutcrackers. However, this unusual and significant behavior was not noted until some years after the study had begun, since normally the females are the most proficient nut crackers, and they were also the most difficult to observe.
47. The significance of this can be seen in the short discussion of methodology included in an edited collection of primate behavior studies that appeared in 1987. Like the earlier volumes it replaced, it was intended to illustrate best practice in current behavioral research, but it echoes their conclusion on the question of experimental practice in the field, arguing that while it was theoretically possible to carry out field experiments without permanently changing the behavior of the animals, primatologists have tended to avoid the risk, either for logistical reasons or because they fear that "behaviour will quickly be distorted if the animals interact with their observers in any way" (Cheney et al. 1987, 7). Instead, their review of field methodology concentrates overwhelmingly on the development and deployment of various sampling methods.

CHAPTER TWO

1. I will discuss "trust" in observers' accuracy and reliability in more detail in chapter 6.
2. Chapter 7 will explore some of the wider consequences of this developing situation in the context of "making sense" of these varied reports.
3. It was in May 1930 that Zuckerman carried out the brief field trip that provided a key part of the empirical basis for his theory of sex and primate society, and it was in 1974 that Hrdy published her first paper proposing that infanticide should be treated as an adaptive product of sexual selection.
4. For example, the intricate social relationships, hierarchies, and lineage groups that are characteristic of elephants are carried on between females. Elephant "society"

consists of matriarchal lineages, and males are found with females only when a female is in heat. For descriptions of elephant society and behavior that bear comparison with those published for the apes and monkeys, see, for example, Douglas-Hamilton and Douglas-Hamilton (1975) and Moss (1988).

5. Eisenberg and his colleagues criticized the overenthusiastic use of the "multimale" troop label, suggesting that most troops so described were not multimale in the sense that more than one mature adult was present, but that the troop instead contained one fully adult male and a number of subadults of varying ages. In such situations, dominance status, for example, might be expected to be a product of age rather than individual capacity. They suggested that the rush to identify troops as multimale reflected a desire to treat primates as "different" from all other animals. The appropriate definition of a multimale troop sparked a debate that continued for some time (Henzi 1988).
6. As a boy growing up in South Africa, he had spent many hours observing baboons (Zuckerman 1981/1932).
7. As I noted briefly in chapter 1, Zuckerman's influence stretched far beyond the anatomy laboratory, not only into the field but also into the corridors of power. He was the first chief scientific adviser to the United Kingdom government (appointed in 1964) and played a key role in British scientific policy for many years. See Peyton (2001) and Burt (2006) for further details. Note also that Zuckerman did acknowledge that grooming and the "hair fetish" might play a role in cementing social bonds between *individual* primates, if not in generating social relationships themselves.
8. See also Chance (1955, 1956). M. R. A. Chance worked in pharmacology at Birmingham University and was, among many other things, responsible for developing species-appropriate housing for nonhuman primates.
9. In relation to the points made later in the chapter with regard to the relation between sex and social rank, Sahlins also emphasized that dominance hierarchies, in the human heritage, had a central role in mediating the problem of sexual attraction.
10. A breeding season, or a "birth peak," would mean that sexual activity, as in the "lower mammals," was limited to certain parts of the year, thus ruling Zuckerman's thesis inadmissible. Zuckerman was extremely affronted by what he considered the misinterpretation of his conception of the "permanent sexuality" of primate groups, and he included his vigorous dismissal of the work of Lancaster and Lee—and by extension, Irven DeVore—in the appendixes to the second edition of *The Social Life of Monkeys and Apes* (1981/1932).
11. See also Collias 1951, 403–4.
12. Washburn (1951) outlines his vision for this new orientation, arguably his most influential paper (Marks 2000). The origins of this framework and its links with the emergence of the "man the hunter" paradigm are discussed in Haraway (1990). Its specific impact on anthropology is assessed in Strum, Lindberg, and Hamburg (1999), an edited collection intended to illustrate the immense influence Washburn had had on twentieth-century American anthropological practice.
13. This troop movement was one of the most memorable scenes in the film *Baboon Behaviour* that Washburn and DeVore made, influencing several generations of primatologists and anthropologists in the United States (Jolly 2000; Rowell 2000).
14. This model was not confined to Western primate observers but was echoed by the work done on Japanese macaques. Imanishi, for example, described the way the

"class organisation and spatial distribution of classes attract the observer's notice" (1960, 398), with the central males and females being surrounded by peripheral animals, drawing on Carpenter's work with rhesus monkeys for support and again emphasizing the role of dominance ranks or hierarchies in integrating the various subgroupings within a primate "oikia." "Oikia" was the term Imanishi suggested for the smallest unit of animal social life (1960, 397). It was not greeted with any great enthusiasm in the West.

15. For discussions, see Zuckerman 1981/1932; Carpenter 1964, 256, 352; Collias 1951; Collias and Southwick 1952; and Chance and Mead 1953.
16. In presuming that males acted as leaders and defenders of their groups, Washburn and DeVore were continuing a tradition that had characterized primate field reports from the early 1950s. Nolte, for example, noted that three large bonnet macaque males seemed to be "leading the troop while on the move" (1955, 178), and March's report of his failed attempt to observe gorillas in Nigeria nonetheless included local hunters' claim that a particular type of nest is "constructed by the leader of the band, who is thus in a better position to sound an alarm" (1957, 32). Donisthorpe, also searching for gorillas, but in Uganda, saw no evidence of special nests but recorded that when confronted by human beings, the leader male would react with an aggressive display, giving his family time to escape (1958, 214). Bolwig, while denying that gorilla males built special nests for observation (1959b), argued that baboon males did "act as sentinels and warn[ed] the troop by their loud barks when danger approach[ed]" (1959b, 140). Frisch's summary of Japanese primate research asserted that among these macaques, subadult males could be seen "watching for oncoming danger" (1959, 588). Kortlandt's chimpanzees included a male he described as the "tyrannical overlord of the band," that "more often than any of other males . . . acted as a kind of security inspector" (1962, 130–31). In these examples, drawn from a range of primate species, different categories of males are being identified as playing a leadership or sentinel role—from a single adult male through to groups of subadults—but in all cases, the existence of this leadership role, and its being performed by males, is utterly taken for granted. Bolwig, in fact, went so far as to suggest that "respect for authority [of the dominant males] is undoubtedly one of the main factors that holds the troop together" (1959b, 145), giving weaker troop members a feeling of security.
17. This has parallels to the later discussions of the impact that the *threat* of infanticide rather than its occurrence might have on the evolution of primate society: the recognition that simply because an observer hasn't seen something happen doesn't mean that it either doesn't exist or isn't important.
18. In other words, that the beta male would theoretically exhibit the second largest degree of mating success, the gamma male the third, and so on.
19. In this context, it is worth noting that at least one reviewer of Schaller's monograph *The Mountain Gorilla* expressed disappointment that "these magnificent, dramatic beasts should lead such dull, ruminant lives" (Goldschmidt 1965, 297).
20. Dominance also needed to be separated from leadership, since from Hall's observations of patas monkeys it appeared that while the male did defend the group from outside, it was the females who decided the direction of march (Hall 1968c).
21. Many studies used "food tests" to establish a linear dominance hierarchy out of interactions between individuals, producing a situation where an observer would decide that an animal was dominant based on its priority of access to food, and that its dominance allowed it priority of access (Gartlan 1968, 96).

22. See also Itani 1963.
23. Here Southwick is citing the work of Japanese researchers on Koshima Island, whose work with Japanese macaques is made available to a wider Western audience for the first time in this edited collection.
24. Schaller, of course, had at this point studied the behavior of two species of great apes (orangutans and gorillas).
25. The difference in social structure between arboreal and terrestrial primates had been known for some time. Chance (1955) argued that group size correlated with habitat, since larger groups were necessary to combat predation pressure on the open savanna (164). Haddow (1952) and Southwick (1963) both tied the difficulty of generalizing about behavior to the variety of habitats where species were found, and DeVore suggested (1963) that the presumed behavioral differences between Old and New World monkeys produced by phylogeny were in fact due to their different ecological niches: researchers were in fact comparing arboreal and terrestrial animals. Jay (1963) echoed this, stressing the problems in extrapolating from the behavior of the terrestrial baboons and macaques to the more arboreal langurs, and DeVore and Washburn had argued (1964) that the more ground-living a species was, the more aggressive and less specialized it would be.
26. Six years later, in 1972, Eisenberg, Muckenhirn, and Rudran produced an updated account of the relation between ecology and social structure that again focused on the critical role of the males within the group and the relationships among them.
27. But of course we should note that among the key figures in bringing the sociobiological revolution to primatology were Irven DeVore and his graduate students. DeVore, as we saw earlier, was himself one of Washburn's graduate students.
28. Here one thinks primarily of situations where individuals refrain from breeding, sometimes to contribute to rearing the offspring of another group member, as with the social insects and avian "helpers at the nest." Other examples can be found in Wynne-Edwards's account (1962).
29. Both Rowell (1967) and Struhsaker (1967) entered the caveat that apparent variability might rather be the product of methodological differences between researchers. Chalmers (1968a) argued that the seemingly clear-cut division between arboreal and terrestrial monkeys was based on unrepresentative samples of behavior in these habitats: it would be necessary to distinguish between *kinds* of behavior (social behavior as opposed to range use, for example) in order to carry out a systematic comparison. Struhsaker (1969) pointed out that most of the attempts to link ecology and social structure depended on savanna or desert cercopithecines, and that including the hard-to-study forest-dwelling animals produced rather different results. His work found that the "direct relationship between group size and degree of terrestriality was not supported" (113) and that there were combinations of variables that did not fit the pattern produced by Crook and Gartlan, for example. He argued that ecology and social structure alone tell only part of the story, since "each species brings a different phylogenetic heritage into a particular ecological scheme" (11)
30. In the sense that it was intended to stand as a general introductory text and sum up the field.

CHAPTER THREE

1. Notably, however, the questions Jay raised about the reliability of the early reports—the problem of anthropomorphism, the quality and duration of observation, the

role of aggression, and the notion of the "typical" monkey—were to recur again and again during the debate surrounding infanticide in the nonhuman primates.

2. Dolhinow later followed Washburn to the University of California at Berkeley, and she remained closely connected with him. It was at Berkeley that she established the only captive colony of langurs in the United States. Her work focused on these captive langurs—rather than on further field explorations—until she retired in the 1990s.
3. This langur group was the only one at Kaukori in 1959. When Kaukori was resurveyed in 1963 an additional langur group was found in the area, and Jay visited Dharwar in 1963 to make short observations there (Jay 1965, 198, 201). From November 1958 till November 1959, she continued observations at Orcha, but they were hampered by the limited visibility at the forested site and the concomitant difficulty of accustoming the langurs to her presence. For the first three months the langurs showed alarm when she came near, but if she "made no sudden movements and followed them whenever they moved, all troops eventually became used to [her] and no longer walked or ran away" (Jay 1962, 285). Her standard practice was to find the group each morning and remain with them as much as possible throughout the day, recording events as they occurred and noting the animals participating and the time of day. She used similar methods at Kaukori from December 1959 to March 1960, although binoculars were less necessary (Jay 1963a, 11) because the langurs were already accustomed to people and the "group could be located and followed at any time of day" (Jay 1965, 201).
4. Although observing animal behavior at Orcha was difficult, it was not impossible. Jay could discern no major difference between behavior at the sites that could not be explained by the relative distances involved. Therefore she chose to use Kaukori as an opportunity to make detailed observations of social activity and Orcha as the site where langur ecology and the nature of interactions between langur groups could be more profitably explored (Jay 1963b, 1965).
5. For example, she shows that the lack of an obvious dominance hierarchy may be related to the way infants are transferred from female to female within moments of birth, and she demonstrates that male and female juveniles begin to show differences in behavior from the age of three or four months, with the males beginning to orient more to the adult males and to use a specific set of signals to attract the adults' attention.
6. So, for example, her chapter headings run in this order: chapter 2, "Infant 1"; chapter 3, "Infant 2"; chapter 4, "Juvenile"; chapter 5, "Subadult"; chapter 6, "Adult Female"; chapter 7, "Adult Male."
7. So, for example, mother, infant, peer, playmate, sexual partner, grooming partner, and so on.
8. Although this journal was originally published in Japanese, it was decided two years later to use English to promote international cooperation. However, it took some time before the journal reached full international standing, since non-Japanese researchers were largely unaware of the primate work carried out in Japan until the publication of Altmann's edited collection of translations of key papers (1967) and the more general introductions provided by Frisch (1959) and Simonds (1962). However, even after this, Westerners continued to regard work published in *Primates* with some suspicion (Asquith 2000, 171–74), which may explain why it took some time for the startling observations of the Japan-India Joint Project in Primates Investigation to be fully assimilated by the nascent international primatological community.

9. See, for example, Haraway (1990), Strum and Fedigan (2000), and Asquith (1994, 2000).
10. The Project involved the collaboration of Indian scientific and educational institutions with researchers from the University of Kyoto and the Japan Monkey Centre. It was funded by a Rockefeller Foundation grant, and—besides Denzaburo Miyadi, included Syunzo Kawamura, Kenji Yoshiba, Yukimaru Sugiyama, and M. D. Parthasarathy.
11. For the reports of the project, see Sugiyama 1964, 1965a, 1965b, 1966, 1967; Sugiyama, Yoshiba, and Parthasarathy 1965; and Yoshiba 1968.
12. Like Jay's langurs, those in cultivated fields were more tolerant of inspection.
13. Within the primatological literature at this point, "troop" is usually reserved for congregations of animals including both sexes and a range of ages; "group" refers to assemblies of males only. I will follow this convention.
14. At this stage of primatological field research, manipulating social structure was not as rare as it was shortly to become even among Western researchers. Additionally, in the case of Japanese primatological communities, human interventions in primate lives did not pose the problem it did within the Western example, not least because the demarcation between humans and animals was less carefully policed than for those influenced by the Judeo-Christian tradition (Haraway 1990; Asquith 1997).
15. The second report dealt largely with langur ecology (Sugiyama, Yoshiba, and Parthasarathy 1965).
16. In the following account, animals are identified by combinations of letters. Capital letters are used for adult males and females, while infants and juveniles are identified by the same letters as their mothers with the letters *I* or *J* following for males and *i* or *j* for females.
17. For example, he began to attack the other males and other potential threats to the troop, and by day five he had successfully copulated with one of the females. However, fighting between the males and particularly the old leader Z did not stop until June 14.
18. "Dogs in the neighbourhood are all kept by villagers. They are not so powerful as to make an attack on the langurs resting on the ground. It is impossible to suppose that carnivorous animals like tigers or leopards left an infant langur only 7 months old half killed" (Sugiyama 1966, 50).
19. KI was abandoned by his mother and as with Ui, his body could not be found when the troop was met the next day, on June 29. By this time Si had disappeared, and "*RI* had been given a serious injury. . . . Bodily injury and killing on *Si* and *RI* seemed to have been committed that morning" (Sugiyama, 1966, 51).
20. Shortly after Sugiyama had found the troop, they met the third troop, but relations between the two seemed calm. Soon thereafter the fourth troop appeared, and at the approach of the leader male the "2nd troop was thrown into confusion and ran away helter-skelter. . . . *IV*'s attack was concentrated on *R*. She ran from place to place with the seriously wounded *RI* under her body" (Sugiyama 1966, 51). The other females surrounded him, with some presenting sexually to him, and *R* with her injured infant kept her distance.
21. The missing infants and juveniles could not have survived alone, since all were still dependent—to a greater or lesser degree—on maternal care. Their deaths therefore could be safely assumed, but the *cause* of death could not be determined.
22. The article consists of fourteen pages of text, seven devoted to the instances of social change the Japan-India Joint Project observed.

23. He outlines three ways the one-male troop structure could be maintained: through the attacks of the male group; by males' leaving voluntarily; or by the leader's active unwillingness to tolerate subadult males. The events observed in the thirtieth troop, involving not only male replacement but male replacement at a point where a troop "that had nearly developed into a multi-male troop was rejuvenated as a one-male troop without having juvenile males or infant males" (Sugiyama 1965b, 401), led him to conclude that the first explanation had been substantiated.
24. This explanation for primate infanticide is still current in modern debates; see, for example, Harcourt and Greenberg (2001).
25. The question of population density, which was to be so crucial later in the debate surrounding whether the tendency toward infanticide is produced by sexual selection, is also raised in these initial reports of infant biting. Sugiyama explains that the reason social change is so common at Dharwar is that langur population density is relatively high there, compared with the animals Jay observed or those surveyed by the Japan-India Joint Project in Raipur. Where population density is high, males that are forced out have no space to form new troops, so in addition to a relative increase in animals in male groups, the male groups must coexist within the home ranges of bisexual troops. These circumstances increase contacts between large male groups and bisexual troops, leading to more frequent social changes (Sugiyama 1966, 70–71). Where the population density is lower, the "opportunity of fighting will be fewer" (1966, 71) and social change—and hence, infant biting—will occur less often.
26. This volume was intended to update DeVore's 1965 edited summation of the state of primatological knowledge. That it needed to be updated so soon demonstrated two things: first, that primatological knowledge and field sites were expanding swiftly, and second, that a desire for community unity lay behind the efforts to collect and collate such information.
27. The sites also showed differences in adult males' responsiveness to infants, although Yoshiba does point out that if Jay had tried to handle langur infants, as the Dharwar researchers did, she might also have seen a protective reaction from the adult males rather than the indifference she argued was characteristic.
28. In fact, in this case Jay's langurs are singled out as unusual with regard to the time spent on the ground and the size of their home ranges.
29. He had observed that in the male groups males lived together without notably overt antagonism (Poirier 1969, 26).
30. He related this to the quality of the maternal care found at the different sites (Poirier 1969, 39–40). Note also his dominant concern with the nature and quality of socialization—in that ecology may cause variation, but so will upbringing.
31. The very title *Primate Patterns* emphasizes the need to appreciate primate variability—within a context of stability. There is more than one pattern of primate behavior, but regularities can be recognized.
32. This suggestion was put forward in a paper included in Poirier's *Primate Socialisation* that discussed the role social organization might play in incest avoidance in nonhuman primates. Essentially, he argued that the one-male troop pattern excluded the possibility of mother-son incest, since males do not rejoin their birth troops, and that the regular replacement of the leader male lessened the risk of father-daughter incest. With regard to infanticide, he raised the possibility that the attack by the new male on the group's youngest infants would eradicate precisely those individuals most likely to have been born as a result of father-daughter incest, since the previ-

ous male's offspring might by that time have reached sexual maturity. In this case the "probability of incest avoidance among Hanuman langurs is greater as a result of this phenomenon" than in other primates—although it does require "the sacrifice of all the infants occurring once every few years" (Itani 1972, 168).

33. By 1974 "infanticide" was already in use to describe attacks on infants.
34. In the sense that it requires animals to act for "the good of the group" rather than in their individual self-interest.

CHAPTER FOUR

1. For example, Robert Trivers, who was central to the development of such fundamental sociobiological concepts as reciprocal altruism, parental investment, and parent/offspring conflict, was one of Hrdy's tutors in her first years of graduate school. He was consistently acknowledged as a major influence behind her work, as were Irven DeVore and E. O. Wilson (Hrdy 1974, 1977a).
2. That is, the reports from Dharwar (Sugiyama 1964, 1965b, 1966, 1967), Jodhpur (Mohnot 1971), and—after Hrdy's studies had begun—Polonnaruwa (Rudran 1973). Ceylon, having achieved independence from the British in 1948, was renamed Sri Lanka in 1972.
3. Calhoun's (1962a) *Scientific American* article directly related social pathology (and especially infant mortality) to high population density. The behavioral repertoire that generations of evolution had bequeathed to the Norway rat could not survive the artificial stress that humans could create in the laboratory. The parallels with the apparent situation at Dharwar were striking, but even more interesting is Hrdy's appeal, at a very early stage of the controversy, to comparative, cross-taxa evidence in the interpretation of primate behavior. As this chapter will make clear, the comparative database was to play a central, though contested, part in the development of the controversy.
4. In this context, "continuous sexual receptivity" means the ability of human females to mate at all stages of the menstrual cycle and even while pregnant. It is rather different from Zuckerman's use of the phrase (1981/1932; see chapter 2 above) to describe a situation in which at least one female in any given primate group would be at the "mating" stage of the estral cycle at any given time.
5. See, for example, Segerstrale (2000) for a history of the sociobiology debates, and also the references in chapter 1, note 5.
6. Originating in the field of engineering, "black box" referred to a piece of equipment so well understood that only inputs and outputs needed to be considered, while the internal workings of the system were unimportant. Adopted by the history and sociology of science, the term now refers to anything—a tool, a strategy, a theory—that has become so fundamental to the operation of a discipline that the costs of challenging how it works are considerable.
7. These troops were identified as Bazaar, Hillside, Toad Rock, School, IPS, and Arbuda Devi—otherwise known as B-3, B-6, B-5, B-9, B-4, and B-2, depending on the audience Hrdy was writing for.
8. The other two took place in the Toad Rock and IPS troops. IPS fissioned during the research, and a takeover occurred in the Toad Rock troop. However, because of the time that elapsed between observations, "it was not known how many infants had been born and if any were missing" (Hrdy 1977a, 266), although the new male in Toad Rock was seen to attack and wound infants. In neither case did researchers know enough about the circumstances surrounding these events to draw any conclusions.

9. Respectively, articles in *Folia Primatologica* and *American Scientist* and her PhD dissertation, which was published as a book. There are certain differences in the way these publications describe these events, not least because Hrdy took the opportunity to include more emotional context in her book-length publication, but there is no disagreement about the events reported at Abu. The biggest difference between the *American Scientist* article and the other two publications is the way it depicts infanticide as a primatewide reproductive strategy rather than confining most attention to the langur literature.
10. United States–India relations were particularly strained during the early 1970s as a result of India's determination to develop both nuclear energy and a nuclear weapons program. It exploded its first nuclear device in 1974, seriously damaging its relationship with the United States and eventually leading to closer international controls on the circulation of nuclear knowledge.
11. As later sections of this chapter will show, the willingness (or unwillingness) of the Indian government to grant research permits was to become a crucial limiting factor in the infanticide debates, especially given that opposition to the sexual selection hypothesis focused overwhelmingly on the langur data.
12. Hrdy notes (1977a, 63) not only that langurs were attacked to drive them away from gardens and rooftops, but that "stoning langurs is a public pastime for young boys and others."
13. However, this also meant she had to draw on the work of Mohnot and Sugiyama for langur ecology because she could not herself produce reliable data.
14. Compare with, "Unusual occurrences in the lives of the langurs were often witnessed by humans. Where reports from reliable informants coincided with what I knew about troop locations and changes in troop composition, these reports were accepted by me as facts" (Hrdy 1977a, 61). In the account oriented to a wider audience, "sometimes" has become "often."
15. "Pat hand," according to Hrdy, refers to Anaconda poker, where players can choose to go high or low. In this game it is possible to draw an unbeatable run of cards, a pat hand.
16. Note that later the idea that only leader males have the opportunity to mate is dropped, particularly after researchers identified multiple matings with extratroop males as a key anti-infanticide strategy for threatened females: if paternity is confused, their infants may survive. But at the moment we are concerned with Hrdy's initial presentation of her hypothesis.
17. So, for example, rather than a male invasion prompting male replacements at Abu, many infant attacks occurred when Shifty and Mug—both troop males in their own right—competed for access to Bazaar and Hillside. A male group invasion did take place in spring 1973, but none of its members were seen to replace either Shifty or Mug (although they attacked infants) until the field season in spring 1975, almost two years later.
18. Hrdy does raise the question of the comparative data in this context, but very briefly, mentioning the "isolated examples" of males' killing infants that exist for other primate species and using Schaller's work on lions to demonstrate that sexually selected infanticide is to be expected under particular social conditions.
19. But see note 14 above: in the article in *Folia Primatologica* Hrdy states that unusual events in the lives of the langurs were "sometimes" witnessed by humans—in the *American Scientist* article it becomes "often."
20. There were fifteen reported takeovers, only nine of which also included infants' disappearances and presumed death.

21. They are "early workers," and their research was carried out when primatology was a "fledgling science" (Hrdy 1977b, 41).
22. Myers (1990) discusses the significance of proposing new research projects in the conduct and resolution of controversy. See also the introduction.
23. Males of the Kasekela community had twice been seen to eat the infants of strange females (Bygott 1972), but the infant-deprived females neither mated with the males nor transferred to their community. Additionally, as I described in the introduction, two females of the Kasekela community had killed and eaten infants belonging to their own community. In neither case did the conditions of the sexual selection model seem to fit.
24. In fact, her letter was titled "Normal Monkeys?" and in the column and a half of print it takes up, "normal" is used four times and "natural" twice (Dolhinow 1977).
25. Notably, she quotes Charles Southwick to demonstrate that the behavior of rhesus monkeys in towns and in rural areas (more "natural" conditions) differed significantly. It isn't clear which article she is referring to here, but bear in mind that Southwick and his colleagues had argued in at least one article that rhesus monkeys have lived commensally with humans in India for so long that it is reasonable to treat urban conditions as part of the evolutionary landscape for groups of this species (Southwick, Beg, and Siddiqi 1965).
26. Her other points cut to the heart of Hrdy's problems with her data set, pointing out that only once had any researcher actually *seen* an infant killed: the rest of the data Hrdy drew on, she asserts, consists of "impressions and hearsay and old anecdotes" (Dolhinow 1977, 266), not least because of the limited time Hrdy was able to spend in the field. Finally, she hints that Hrdy had anthropomorphic tendencies, since she attributed "incredible powers of memory and reason" to the animals (1977, 266).
27. Where Hrdy uses the term, she consistently places it in scare quotes, showing she is unhappy with the concept.
28. Robin Dunbar described this pattern in relation to the assumption that there was, for primates, such a thing as "species typical" behavior. He argued that while the sociobiological revolution had largely eradicated this tendency for animal behavior in general, "primatologists carried on blithely regardless, sometimes in the most surprising places, and you have these extraordinary arguments going on in the seventies, late seventies, even early eighties. . . . If you've got two populations behaving differently, then you had one [researcher] saying, well, your population is disturbed, it doesn't count, they are behaving atypically."
29. In this context, consider the subtitles used in the two *American Scientist* articles. Hrdy's rejection of the pathology hypothesis in favor of an evolutionary explanation rooted in conflict between individuals is signaled by the first word of her subtitle ("Conflict Is Basic to All Creatures That Reproduce Sexually"). In contrast, Curtin and Dolhinow begin their attempt to reinterpret the events in India for a Western audience when their own subtitle says, "Human Alteration of the Environment May Be Pushing the Gray Langur Monkey of India Beyond the Limits of Its Adaptability." This article was also published the next year in *Science Today,* an Indian popular science journal run by the Times of India group. The only major change to the article was in its title, which became "Infanticide among Langurs—a Solution to Overcrowding?"
30. The history of science is filled with instances where the veracity and accuracy of observations by nonscientists were challenged, not least because there was little cost to

disbelieving their words. See, for example, Shapin (1994), Duncan (1987), Browne (1996), McCook (1996), and Carey (1997).

31. Curtin (1977) puts forward much the same argument but in far greater detail.
32. However, at this point Hrdy was, along with other colleagues, trying to develop a schema by which the relative levels of disturbance at a site (human influence, harassment by predators, and so on) could be judged with some detachment (Bishop et al. 1981).
33. Here we have a clear parallel with Collins's "experimenter's regress"—with the field site or environment itself being the functional equivalent of the experimental apparatus in the gravitational waves controversy, for example.
34. In fact, Bishop (1979) publishes an account of the behavioral pattern of the Himalayan langurs that suggests they are both "morphologically and behaviourally distinct from conspecifics in other ecological niches" (279), with a cluster of behavioral features that, appearing together, are unique to these groups.
35. It was true, however, that experiments run at the Berkeley colony resulted in infant death. Dolhinow and her colleagues (Dolhinow 1980; Curtin 1981, 1982) describe events surrounding the experimental investigation of how losing their mothers affects infants. Of twelve infants separated from their mothers after weaning, two died.
36. In the same year, Mildred Dickeman (1975) bemoaned the relative neglect of infanticide by social scientists and argued that, unlike animal infanticide, human infanticide was normally deliberative and selective. She quoted another researcher as suggesting that part of the definition of man might be "the only animal to purposefully commit infanticide" (1975, 108).
37. She begins the section devoted to sexual selection with a review of the experimental work on lemmings that seems to fit with the predictions of the hypothesis, but most of this discussion concentrates on the nonhuman primates, although she does briefly mention lions.
38. Partly this may be because Boggess published in *Folia Primatologica,* a journal specifically concerned with the primates, whereas Hrdy's work appeared in *Ethology and Sociobiology,* which had a much broader mandate. However, this does not explain why Boggess did not discuss the comparative *primate* literature, which was by now playing a major role in the debate.
39. For Dharwar it was assumed that the new male resident in the troop had killed the infants, but only from circumstantial evidence, and the paternity of both the killed infant and subsequent infants was unknown. For Jodhpur, the only case of infant killing was ascribed by the observer (Mohnot) to the high levels of aggression, and in any case, not only did the new resident male not copulate with the infant-deprived females, he killed the offspring of another female he had mated with. The data from Abu are extensively critiqued, with particular attention to the lack of direct observations of infant attacks. Boggess rejects most of Hrdy's observations with her conclusion that "as most of the infanticide data from Abu result from temporal correlations between male membership change and disappearance of infants, rather than direct observation, the precise causal relationship between the two events is difficult to establish" (1979, 77–81).
40. "Allomothering" occurs when females other than the mother take care of an infant.
41. For the behavior to be useful to the infant, the allomother should be an experienced adult female—who not only would not benefit but would be harmed by wasting time caring for another female's infant: for the care to benefit the allomother,

she should be inexperienced and using the opportunity to become familiar with infants—which would be positively dangerous for the infant. They suggested instead that allomothering should be considered a by-product of a more general selective pressure for female interest in infants—and as a demonstration that even common behavioral patterns may not be adaptive in themselves.

42. Handling another female's infant would let females learn about that infant but could harm its chances of survival, not least by depriving it of its mother's milk.
43. It's notable here that Packer does not refer to the controversy over infanticide at all.
44. Again, in relation to the later development of the controversy, note that Makwana tabulates assaults, disappearances, and deaths. Except for two adult females who died from electric shock or dog attacks, all other injuries or disappearances are followed by the remark, "Supposed to be injured by . . . " one or other of the males (1979, 297). Clear observations of attacks are again missing, and on at least one occasion, Makwana's knowledge of what occurred among the langurs is based on information from local people (296).
45. More indirectly, Packer (1980) and Busse and Hamilton (1981) suggested that a well-known strategy of male baboons—carrying infants when tension between males is high—might be directly related to infanticide and to reduction of the infanticide risk. Previously it had been suggested that males carried their opponents' infants to reduce attacks, since their opponents would be unlikely to risk damaging their own offspring. Instead, these researchers argued that males carried their own infants to protect them against any infanticidal tendencies of an incoming male—parental care that explained why infanticide reports are rare for baboons.
46. This meeting was held before the Eighth Congress of the International Primatological Society, which met in Florence that year. Attending the symposium were Christian Vogel, Axel Goldau, Jutta Kuster, Hartmut Loch, and Paul Winkler (University of Göttingen); Silvana Borgognini-Tarli (University of Pisa); Glenn Hausfater (Cornell University); Anne Baker-Dittus and Wolfgang Dittus (Smithsonian Institution Primate Project); Sarah Hrdy and James Moore (Harvard University); and John Oates (Hunter College).
47. Vogel worked at Jodhpur, where no infanticides had been seen since Mohnot's observations in the early 1970s.
48. Note that infanticide was not the *only* topic discussed at this symposium—participants also touched on female counterstrategies to infanticide, allomothering and infant abuse, and aging and social rank in both sexes. Infanticide, however, was the main focus of debate.
49. Although, having made this point, they go on to grant that the supposition that the new male was to blame was most likely true.
50. This is related both to Hausfater's letter to *American Scientist*—that neither Dolhinow nor Hrdy should treat their sites as necessarily representative—and to the later attempts by Sussman and his colleagues to argue that since infanticides have mostly been observed at Jodhpur, then infanticide is a result of the oddities seen mostly at Jodhpur. See chapter 5 for a detailed discussion of this point.
51. Schubert was also a member of the International Primatological Society and of the American Society of Primatologists. After devoting the first part of his career to the study of judicial behavior, in the late 1970s he developed an interest in the biological roots of political behavior and eventually published two books on the topic.
52. He points out that six of the infant disappearances took place in the first month of the study, "when she was herself necessarily a somewhat naïve and in any case a

very inexperienced observer of feral langurs" (Schubert 1982, 212). He draws attention to what he considers contradictions in her account of the ages and coloration of the infants as Abu and emphasizes that the total number of infants considered to have been killed is not consistent even within Hrdy's own account—for example, in the initial case of six missing infants, Hrdy first ascribes only two infant deaths to the new male, but by the conclusion of the article she includes all six in the tabulation of infanticides.

53. Hrdy, personal communication, 4/5/02.
54. Naturally she also reacted emphatically to his attack on her professionalism, reminding the audience that she went to the field not to prove the sexual selection hypothesis but to test the idea that social crowding was responsible for attacks on infants. She also tried to correct some misinterpretations of her data that she identifies in his article.
55. In Hrdy's own fieldwork, locals twice attacked her for spraying langurs with colored dye to identify individuals—they thought she was selecting them for slaughter.
56. These three consisted of Boggess's original paper and the two that Sugiyama and Hrdy produced as commentaries on her paper. However, while the commentaries on other papers published in this section had been incorporated before going to print, it was impossible to achieve this for the langur controversy.

CHAPTER FIVE

1. Again, note that there are at least four "adaptive" models for infanticide in general (Hrdy 1979): the sexual selection model is only one of these, although it deals with the most commonly reported type of primate infanticide.
2. On predictions: there are several predictions for the adaptive model, some proposed by opponents, some by supporters. It would take too much detail to outline them here, and in any case such a discussion is best left to later chapters. I have included the predictions that seem the most basic and also common to all the reported primate infanticides.
3. The German and Indian teams used different names for the troops.
4. Carola Borries received her PhD from the University of Göttingen and is now at the Department of Anthropology at the State University of New York at Stony Brook. With her partner, Andreas Koenig, who is also based at Stony Brook when in the United States, she has spent the past fifteen years working on primate behavior with both captive and free-living animals.
5. Additionally, researchers there were able to study the male groups surrounding the bisexual troops and discovered not only that it was almost impossible for any male other than the troop leader to father infants (Sommer and Rajpurohit 1989), but that males seemed to time their attacks to increase the likelihood of immediate reproduction (Winkler 1988).
6. Almost half of the infants in a group at the time of a male replacement were not attacked by the new male: of those born afterward, more than half escaped unscathed.
7. He notes here that some disaster may well have struck the low-density population before the study began: the population steadily increased (despite the infanticides) over the course of the study. He argued that this increase demonstrates the danger of treating cross-sectional studies as representative of the population, echoing Rowell's (1967) warning about the dangers inherent in treating the present state of study sites as representative of their historical status.

8. The collection also devotes a chapter to infanticide, which reports new cases but follows what is now a common pattern: it acknowledges a relative dearth of directly witnessed accounts but emphasizes that what was seen of each interaction fits more closely with the predictions of the sexual selection model than with any other hypothesis (Struhsaker and Leland 1987).
9. Anne Pusey gave the following account of how the filming proceeded: "Alan Root . . . decided that he would make a lion film; he wouldn't make it [himself], but he'd send out a film team to make it. So a couple arrived and spent eighteen months in the Serengeti making the film. At the beginning we talked a lot with Alan and the film team about what would make a good lion film, and we said, you know, you have to film infanticide because it has such an important effect on the whole society. So we said, it happens when the female's not well defended, and we have a couple of females here at the moment, with collars on, and they're about to have cubs, or they've had cubs, and if you go out and watch them, you'll see their cubs being killed. We'd never bothered to do that ourselves because we knew it happened all the time, and it's hard to see it happen because it's usually at night. So [the film crew] watched the first female and then one night, they saw it, but it was dark so they couldn't film it. So then they went to the next female and they watched her for days and days and days, and we didn't know how possible it would be to see it, and then they [managed to film it] . . . in the middle of the day! . . . The unedited film is horrific, you know, he goes from one cub to the next, to the next, picks it up, it's very clinical, he kind of picks them up, shakes them, bites them, and then goes on to the next one."
10. Clearly, this doesn't entirely explain the evolution of lion sociality, since infanticide had also been observed in other, solitary, big cats (Smith and McDougal 1991).
11. Hausfater and Hrdy 1984; Parmigiani and Vom Saal 1994.
12. Females have been cited as responsible for adaptive infanticide in dwarf mongooses and prairie dogs, some domestic house mice and all wild mice, and some social insects—but on some occasions a female will parent another female's offspring.
13. The "Sokal affair," generally considered one of the most famous events of the "science wars," occurred a scant two years after this book was published (Sokal 1996a, 1996b); see my introduction, note 4. It is reasonable to assume that hostility toward what appeared to be the mounting strength of an "antiscience" movement in the academy was already growing.
14. Having received his PhD in anthropology from Duke University, Sussman had already spent many years in primate fieldwork. He has published over a hundred articles on primate behavior and ecology and most recently produced a three-volume account of the relation between primate ecology and social structure (Sussman 2003).
15. This complaint is particularly interesting given Hausfater and Vogel's call (1982) to include just this sort of detail in reports of infanticide. Presumably, by this stage the controversy had been quiescent long enough for such discipline to appear unnecessary.
16. Given the difficulties of measuring fitness and success in the field, researchers often used a shorthand for assessing whether the male was likely to have benefited from the infanticidal attack: measuring the length of the interbirth interval and comparing it with a "normal" interbirth interval where the infant survived. If the former was shorter, they concluded that the male had increased his fitness. Sussman and his colleagues were insisting that this measure was inappropriate because it ignored

the fact that infants die for many reasons (predators, disease, accident, drought or famine). Male "fitness," they argued, can be safely measured only by putting suspected infanticides in the context of overall mortality.

17. John Fleagle was the editor of *Evolutionary Anthropology,* the journal in question.
18. This can be related back to Hausfater and Vogel's points described in chapter 4: simply counting the number of infanticides reported is not an appropriate way to assess the validity of the hypothesis: different species and different ecologies will show clear variation in the extent and frequency with which they meet the conditions laid out by the hypothesis.
19. Most examples of predation in the literature are presumed rather than witnessed (sudden animal disappearance, for example, or finding monkey parts in predator feces, as reported by Phyllis Dolhinow). Despite this, predation had long been cited as a key factor in the evolution of primate societies, and the lack of witnessed episodes was made up for by recognizable antipredator mechanisms (alarm calls, defensible sleeping places, and occasionally cooperative defense). Similarly, Hrdy and her colleagues argue that there is a range of animal behavior that makes sense only in relation to the threat of infanticide, such as paternal behavior, infant abandonment, or sudden changes in mothers' ranging patterns. However, all these examples could plausibly have other, unconnected explanations: only if the audience accepts the sexual selection model can they be seen as part of a wider behavioral complex intended to alleviate infanticide risk. The parallels with the "experimenter's regress" are striking.
20. Sussman and his colleagues' reasons for excluding the comparative literature will be explored in detail toward the end of this chapter and in chapter 7.
21. Sussman (2000) quoted it in his own account of the continuing controversy in *Primate Encounters,* and Dolhinow (1999) cited it as an admission by supporters of the sexual selection model that their data were inadequate.
22. In this context, note several points. The first concerns "science by press release," perhaps most strongly associated with the "cold fusion" affair of the late 1980s. The general assumption had always been that if the reliability or validity of their claims was suspect, scientists would circulate their ideas through the media rather than through the peer-reviewed literature. The study of the cold fusion affair instead demonstrated that mass media reports were key to the falsification of Pons and Fleischmann's claims (Lewenstein 1995), against a background of interpretive and reflexive flexibility in testing such claims (Gieryn 1999; Collins and Pinch 1993). Simon's later work (1999, 2002) demonstrated that although ending the controversy also ended "official" study of cold fusion, many reputable scientists continued to work on the problem in their spare time and after hours. In other words, while circulating one's views through the mass media may at face value be taken as demonstrating the weakness of one's position, the charge has not been borne out through further investigation. The second point, however, is that those making novel claims understandably want to publicize their work as widely as possible, especially where it challenges a taken-for-granted claim (Myers 1990). Third, university press offices are now much keener than in the past to take an active role in disseminating research.
23. In these cases animal examples are given as if they are directly relevant to human experience.
24. The invading male coalition must be closely related, it must kill cubs and return mothers swiftly to estrus, it must mate with the deprived mothers, and then it must

stay with that pride for the next two years until the cubs are independent: if the males leave, the next male coalition will kill their cubs.

25. Schaller began work on the lions of the Serengeti in 1966. He was followed by Brian Bertram (1969–74) and by Jeanette Hanby and David Bygott (1974–78) before Craig Packer and Anne Pusey arrived in 1978 (Schaller 1972; Bertram 1975, 1976; Hanby and Bygott 1979; Packer and Pusey 1983).
26. Only seven takeovers had been seen in seven years, and in only four cases were cubs killed: the correlation, she suggests, is extremely weak, since infant death may be linked just as satisfactorily with maternal neglect. There is striking similarity between this process of elimination and the approach that Dolhinow and Sussman and their colleagues take to the primate literature.
27. The letter was not published in *American Anthropologist* but appeared in *Anthropology News* (Silk and Stanford 1999).
28. I will develop this point further in chapter 6.
29. The "four fields" were cultural anthropology, physical anthropology, linguistics, and archaeology. Clearly, an anthropology department that covered all four would include staff drawn from many areas in the natural sciences, the social sciences, and the humanities.
30. Dagg herself admitted, as did Sussman, that her original paper, "Lying about Lions," was roundly rejected when it was first written in the early 1980s. However, the paper was sent to five referees; four recommended publication, but the fifth suggested it be sent to an animal behavior journal (Shea 1999). Naturally, referees are anonymous.
31. Dagg was criticized for rejecting the observations of highly experienced lion watchers when she herself had never systematically observed wild lions. Additionally, her attempt to use the lion data to challenge the credibility of the sexual selection hypothesis without discussing the rest of the comparative literature, including the experimental evidence, was called questionable. In response, she asked whether her critics were each thoroughly experienced observers of rodents, ground squirrels, and so on—they must be, or they could not speak with authority about that literature. But she was willing to discuss the comparative literature in relation to pinnepeds and chimpanzees, where the adaptive function of infanticide was recognized as unclear.
32. This includes the idea that both males and females kill cubs because they want sex—rather similar to Sugiyama's initial hypothesis.
33. She also comments, "Perhaps being continually observed year after year by human fieldworkers itself is stressful for lions?" (Dagg 2000, 833).
34. See Fedigan (1992/1982) and Strum and Fedigan (2000) for a more detailed account of this shift. However, primatologists were not alone in realizing that activities between females were worthy of note—see, for example, Wasser's edited collection (1983) and Wrangham (1997).
35. Females had been seen to kill infants under conditions that suggested they were eliminating a future competitor either for resources (Agoramoorthy 1994; Agrell, Wolff, and Ylönen 1998; Pusey, Williams, and Goodall 1997) or for access to males (Arcadi and Wrangham 1999; Clutton-Brock et al. 1998; Digby 1995, 1999). However, infanticide by female primates was even more rarely reported than infanticide by males, which continued to dominate the literature.
36. Possibly in response to the way Sussman and his colleagues had treated rates of infanticide reporting, she was concerned to emphasize that the poor visibility at

Ramnagar meant the number of reported attacks could not be directly compared with those at other sites where visibility was much better.

37. Note that by now this was the most common explanation for infanticide in chimpanzees. Infanticidal chimp males, it was concluded, were cooperatively defending females' access to resources and thereby protecting and insuring their reproductive investment in the offspring of those females.
38. And indeed, Patterson et al. (1998), for example, agreed that even though actual infant killing had yet to be observed for dolphins, it must still be taken as a fundamental factor in shaping social systems.
39. Dolhinow continued to emphasize that infant attacks had not been observed at every site where langurs were studied. She also continued to imply that treating infanticide as an example of sexual selection required assigning immensely complex mathematical and mnemomic capacities to the male.

CHAPTER SIX

1. As the introduction indicated, all the scientists I interviewed had the opportunity to correct the transcripts of their interviews. They have also been given the chance to review any direct quotations from these transcripts. See the appendix for a list of participants and interview dates.
2. By the middle to late 1990s some scientists and their supporters were responding to what they saw as an attack on science by some historians, sociologists, and philosophers of science. See note 4 in the introduction for a brief account of some of the key publications in the science wars.
3. In "playback" experiments the researcher records the calls of various animals under different conditions, plays them back to the other animals, and monitors their responses. Perhaps the most famous of these experiments were those that Robert Seyfarth and Dorothy Cheney carried out at Amboseli (Cheney and Seyfarth 1990), where they were able to demonstrate not only that vervet monkeys could recognize the calls of individual animals, but also that they used distinct alarm calls for different predators. In the case of infanticide, however, this technique is questionable because of its potential consequences for primate behavior and survival. For example, Alexander Harcourt wondered what might happen "if we got the animals used to a strange male's vocalization and the strange male was infanticidal. Imagine a small population of gorillas in which you get animals to cease to pay attention to strange males? We could have a serious impact on the health of that population." Playback experiments therefore must be conducted with caution.
4. This bears some similarity to a point that came up several times in an earlier project focused on the development of primatological field practice. Several researchers argued that reviewers rejected their papers on the grounds that they had too little evidence to support their arguments and that they were required to collect more examples. As one researcher put it, "But you know, there are only six gorillas" (2001a). A limited number of animals are available for study, and more cannot be added as easily as lab researchers can order more rats.
5. Some researchers also raised the point that primates in general are comparatively long-lived, slow-developing animals. Studying primates is difficult because their life spans can at least approximate the active working career of a field scientist. One's attention is therefore directed toward recording behavior rather than manipulating it.

6. And they might even be used to develop an experimental protocol in captivity, "where you could [test this] without actually having an infant harmed, though you might have a mother pretty alarmed, and the mother's alarm might even be part of your measures." Louise Barrett described similarly characteristic behavior in the epigraph to the introduction.
7. This was, for example, the focus of many of the studies in Van Schaik and Janson's edited collection (2000).
8. As Kelly Stewart said, "Even with mice and rats, many investigators don't allow the infanticide to occur. They stand there and grab the babies away at the last minute. So even there it's considered unethical to create a situation in which babies are killed, deliberately create a situation in which infants are killed."
9. Remember that the only captive langur colony in the United States was under the direction of Phyllis Dolhinow at Berkeley. It may also be worth bearing in mind that studies done there had led to infant death—the mother-separation experiments carried out in the late seventies and early eighties.
10. Sarah Hrdy pointed out that this was not surprising, since sociobiological theory tended to encourage researchers to consider the comparative data when studying specific problems. However, it makes Dagg's attack on the lion data, sponsored by Sussman, rather more interesting: if the comparative evidence is irrelevant to understanding primate infanticide, why should a critique of the lion data be published in an anthropological journal?
11. In fact, several researchers argued that it was only in Van Schaik and Janson's (2000) collection that the cross-species evidence for sexually selected infanticide had been fully integrated.
12. So, for example, Kelly Stewart argued that "it's fair to set up two types of infanticide and look at the data separately": that is, to ensure that it's possible to distinguish between events known in differing degrees of certainty. Similarly, in retrospect Sarah Hrdy would "have made [the 1974] paper longer. And there would have been a whole lot of . . . splitting it up into what was seen; I would have talked about things like, who is this unidentified police cadet, they weren't just guys off the street . . . did you see [the baby] actually attacked, did you see blood, did you see the body later . . . would have been much more paranoid about how I reported that information." In fact, as the first epigraph to this chapter suggested, there was general agreement that the debate had been helpful in forcing people to be far more specific about what they had or had not seen, and under what conditions.
13. This point is important to apprehend. Part of the problem is that two sides are taking very different approaches to how evolution works in practice. For the supporters it is a fixed trait that has in the evolutionary past established certain patterns of social relationships, and they are now interested in understanding how this behavioral complex can help to explain current relationships. For the opponents, they take the position that supporters are trying to suggest that evolution is occurring as we speak, that males that are committing infanticide are leaving more infants that those that are not. For the supporters, it is a condition-dependent behavior—if a male doesn't do it, it doesn't mean he doesn't possess the behavioral complex, but only that he has never found himself in the right situation to express it. It has come to fixation in the population.
14. Nowhere, it was argued, was this clearer than in the Serengeti lion study. This study differed from most primatological examples in that infant killing in situations of

male takeover was close to 100 percent. See chapter 5 for a description of the predictability of lion attacks in the context of filming their activities.

15. That is, to transfer males into groups that habitually contained more than one male.
16. Now retired to Yorkshire, Rowell spent most of her career at the University of California–Berkeley.
17. That is, the behaviors described in the previous section as characteristic of an infanticidal male.
18. Again, this assumes that the supporters of the hypothesis are arguing that evolution is occurring constantly. In contrast, they might argue that they take a slightly different perspective—that current behavior reflects evolved solutions to past problems—which may or may not still be present.
19. Although, as chapter 5 demonstrated, they do have many problems with the quality and extent of these data.
20. Again, this problem arises from the structure of primate field studies: as more is learned about the population rather than the group, such questions can be answered, but learning enough to do this is long, laborious, and expensive task.
21. Identified by Grafen (1982)—a principle that allows you to ignore the unknown genetic mechanism in order to focus on the behavior.
22. Again, the shortening of the interbirth interval is the shorthand that lets researchers say that males have improved their fitness.
23. In relation to this, Robin Dunbar suggested that this confusion was one of the key sources for sociobiological controversies in general, in that people were making category mistakes: "You've had two kinds of argument being deployed, sociobiologists are trying to develop functional explanations in which genes have consequences you can check, and you've got the critics attacking them on grounds that have to do with genes as causes of behavior, which is an ontogenetic argument. It simply isn't the case that the genes that are being talked about in the one [example] are necessarily the genes being talked about in the other. In very simple biological systems, they will be, but in more complex systems, dealing with higher vertebrates, where you have behavior and minds intervening, they aren't."
24. But to tie the occurrence of the controversy to this question would leave open why infanticide in particular became the subject of controversy.
25. And this information is the very minimum you need simply to propose the hypothesis. Testing it is quite another issue—as Jim Cheverud argued, "Some of the concentration on showing individual cases of chains of events doesn't really answer the question," since "you could find a million infant killings and it wouldn't be proof of a model."
26. So, for example, Alexander Harcourt commented about Sussman's work that "those few pages where he's got [it] wrong will be ignored . . . and the rest of it is fine."
27. This is part of the reason the comparative literature was considered less relevant for primatologists.
28. That is to say, it had to accommodate the human perspective where primate studies were or are still found in anthropology departments or divisions. It would not be so where primate research was located in biology or psychology, for example.
29. Some researchers—Daly and Wilson among them—have argued that this is true.
30. Although some interviewees were less scathing about anthropomorphism, suggesting it was less questionable than it used to be—in that it was permissible to use anthropomorphism in hypothesis construction as long as one could avoid it com-

pletely when testing the hypothesis. See also the developing literature on human-animal relations and the question of what animal behavior can tell us about human behavior: among others, Burt (2003), Daston and Mitman (2006), Franklin (1999), Ham and Senior (1997), Haraway (2003), Henninger-Voss (2002), Mitman (1999), Mitchell, Thompson, and Miles (1997), and Crist (1999).

31. Let me reiterate that before infanticide was studied in nonhuman primates, the deliberate killing of young had been suggested as an uniquely human trait (Dickeman 1975).
32. Robin Dunbar pointed out that "the literature is littered with all these metaphorical uses, for better or worse, the kamikaze sperm hypothesis, the Three Musketeers effect, you can just go on and on."
33. Carel van Schaik emphasized the importance of recording "morphological descriptions of the behaviors before you hang functional labels on them" because of the danger of "thinking you've solved the problem" before the functions of that behavior have been thoroughly investigated.
34. Sarah Hrdy, for example, was emphatic that Dolhinow "believed that individuals were behaving in such a way as to enhance the reproductive survival of the group," and Alexander Harcourt described the hostility some Japanese workers expressed to the idea of individual competitiveness, which he also attributed to Dolhinow.
35. As Phyllis Lee put it, "It was interesting, having gone from a perspective on langurs as really sweet, playful, caretaking, then you get this perspective that sexual selection means the males are out there killing the babies all the time. So it was obvious that it was going to raise controversy; people had barely moved out of group selection. The old tradition that this couldn't be happening, it wasn't for the good of the group to kill babies, was still deeply rooted." Recall also that Hausfater had argued (1984) that the infanticide debate marked the disciplinary maturity of primatology, in that all members of the discipline could (potentially) agree on the usefulness of sociobiology.
36. In which members of the social sciences must be included.
37. Several interviewees were tremendously concerned to emphasize how seriously they took this problem, not least because of its consequences for the disciplinary unity of anthropology in the United States as departments fissioned into separate groups of biological and cultural anthropologists. Tim White, based at Berkeley, regretted this, pointing out that historically "anthropology was one of the few disciplines that bridged the social and natural sciences. And that is killed when anthropology disintegrates in the way that it has."
38. This had emerged from the reactions to a proposal to rename the graduate program in anthropology as the "interdepartmental doctoral program in anthropological sciences." Opposition had centered on the use of the word sciences, with several colleagues apparently rejecting the attempt to describe their work as scientific.
39. This quotation comes from Steven Jay Gould's response to the decision of the Kansas Board of Education to delete the teaching of evolution from the state's science curriculum. He was responding to the board's assertion that evolution was not well enough documented to be treated as "factual." Sarah Hrdy uses this quotation as an epigraph in her foreword to Van Schaik and Janson's *Infanticide by Males and Its Implications* (2000).
40. Note that there is clear room for disagreement on this last point. Several supporters of the hypothesis took issue with what they saw as the tendency to overstate the case

for the influence of infanticide on primate societies and believed its potential effects remained to be proved.

CHAPTER SEVEN

1. When one considers the difficulties inherent in doing science in the field, it is surprising that such intransigent controversies are not more frequent for field scientists.
2. So, for example, while most of the long-term, continuously operated field sites were in East Africa, in countries such as Tanzania (Gombe, Mahale, Mikumi), Uganda (Kibale), Rwanda (Karisoke), and Kenya (Amboseli, Gilgil), reports on primate behavior were being made from all over the African continent. For India, sites such as Jodhpur were under relatively constant monitoring, and several Sri Lankan sites were also producing regular reports. Work was going on in Malaysia and Micronesia (Kawabe 1970; Koyama 1971; Southwick and Cadigan 1972; Kawabe and Mano 1972; Poirier and Smith 1974). In Central and South America, besides the long-established captive colonies at Barro Colorado and Cayo Santiago, work had been done on the green monkeys imported into St. Kitts by European slave traders (Poirier 1972), in Panama (Baldwin and Baldwin 1972a, 1972b, 1973), and at the Hatao Masagural ranch in Venezuela (Neville 1972a, 1972b; Oppenheimer and Oppenheimer 1973). This is by no means an exhaustive list, but it should give an idea of the range of primatological field research that had been established by the early 1970s. This list was to grow exponentially over the next two decades.
3. For example, many sociobiological hypotheses can be tested only where individual life histories are known and lifetime reproductive success can be estimated or judged—that is, at sites where long-term data sets exist. Another reason researchers were willing to consider "disturbed" sites was the awareness that adaptability was the hallmark of the primates generally—disturbed sites therefore become places where the limits of adaptability can be assessed.
4. Which in turn rests on what is taken to count as "appropriate" behavior.
5. And could not be known: the ideal length of a field study can literally be "as long as possible": as long as the money holds out, as long as the animals remain visible, and as long as the researcher lives. As Rees (2006a) demonstrates, a "lengthy field study" was defined very differently in the late 1960s and in the late 1980s, and some field studies established in the 1960s are still ongoing.
6. Again, this number is not readily definable. In practice, researchers tended to conduct repeated surveys of the animals, to focus on a few core troops, or to use some combination of these techniques. In either case, "as many as possible" had, de facto, to be defined in context rather than established as an ideal.
7. So, for example, there were some cases where it was clear that animal behavior had been modified by human influence—where animals lived close to towns and villages and mingled freely with humans. Notably, however, in these particular examples researchers found it necessary to argue that they could treat the animals as true exemplars of "natural" behavior. See, for example, Southwick, Beg, and Siddiqi (1965), where the authors argue that the rhesus monkeys under observation have lived close to humankind for so long that this "commensal relationship in villages, towns, temples and roadsides represents a natural relationship" (158).
8. Clearly these two techniques were much easier to use with large semiterrestrial animals living in open country than they were with smaller forest-dwelling or nocturnal primates.

9. DeVore's collection is not the only one to use this technique: it frequently appears in the first descriptions of the social behavior and characteristics of a particular species. See, for example, the other edited textbooks that appeared before the early 1970s and articles in *Folia Primatologica* from 1963 to 1973.
10. That this technique did have this result is demonstrated by the surprise early primatologists expressed about the documentation of primate variability—*the* baboon behavior had already been described from beginning to end—and by the presumption among those I interviewed that belief in species-typical social structure had persisted much longer in primatology than in other ethological disciplines.
11. Washburn and his colleagues by no means discounted the influence of ecology: at the heart of the "baboon model" was the notion that particular environments tended to produce particular social structures (Washburn and DeVore 1961; Haraway 1990; Rees 2006b). They did treat it as less relevant, however, and placed less stress on the problem of variability itself. They suggested, for example, that the appearance of significant variation in primate behavior might simply result from a short-term variation in food supply (Washburn, Jay, and Lancaster 1965).
12. Although, as they admit, it was harder to classify the great apes and humanity itself according to the grades and categories they had identified, they did make some tentative suggestions about the impact the savanna environment might have had on early humans.
13. Although it may be seen as an immediate outgrowth of the calls made in the United States in the 1940s for studying animal behavior as a means both of understanding human behavior and of ameliorating if not eliminating social problems such as crime, overcrowding, and warfare. As I indicated earlier, using animal behavior as a model to account for human actions has a venerable history.
14. Note, however, that Hrdy, and her colleagues, did try to outline parameters for estimating the levels of human disturbance to sites in South Asia (Bishop et al. 1981).
15. Perhaps one of the best examples here is the work of Richard Wrangham, whose approach to the study of behavior is thoroughly sociobiological but who also treats ecology as a fundamental factor (Wrangham 1980). See Janson (2000) for a discussion of the significance and influence of ecological perspectives in the development of primatology.
16. This technique was used several times in the infanticide controversy—see, for example, Newton (1988) and Van Schaik and Dunbar (1990). In fact, Robin Dunbar described Newton's work as "rather clever really, and [something that] isn't done often enough in the behavioral sciences, [which is to] pit the two hypotheses against each other, in effect to test between them using the same data sets. . . . What he does is to say, okay, if population regulation is the issue, you should find all the infanticidal sites over on this side of the graph at the high population densities, and the noninfanticidal langur populations over there, and if it's sexual selection, you should find all the infanticidal populations at the top where they're mostly one-male groups, and noninfanticidal sites at the bottom. So in a sense you're putting to the data themselves the question, Which do you believe? Not the humans; you're asking the data."
17. That is, until Sussman's publication of Dagg's 1999 article, which caused consternation and anger rather than critical debate.
18. Again, *most* circumstances, not *all*. Chapman and Hausfater's mathematical projects, and their later elaborations, laid out the conditions under which males were likely to benefit from infant killing according to the length of the interbirth interval and the period of male control over a reproductive group.

19. Again there is a parallel with the wider case of sociobiology—Richard Lewontin, among others, has made similar demands in his criticisms of sociobiology. I am indebted to an anonymous reviewer for reminding me of this.
20. Or where species were being observed at sites that differed ecologically, it might be that apparent differences in behavior were the product not of different ecologies, but of different conditions of observation imposed by the site's ecology.
21. See, for example, Rees (2006a) for a discussion of these issues specifically in relation to field primatology, which can be put in the wider context of the history of field science literature. (See chapter 1, note 28.)
22. The difficulty of accurately observing fast-moving and stressful events can perhaps be illustrated by one of Hrdy's experiences at Abu, previously described in chapter 5. She used a movie camera to record events so as to extend and reinforce her own observations, and on reviewing some of the infant attacks, she was startled to realize that the langur male Mug, rather than attacking the infant, may well have been defending it from the other males in the group.

CONCLUSION

1. One of the best discussions of the philosophical, ethical, legal, and theoretical questions about the political consequences of science studies is in the special issue of *Social Studies of Science* 26, 2 (1996). Also, see the debates about Collins and Evans's proposal that science studies had entered its "third wave" (*Social Studies of Science* 32 [2002]: 235–96).

REFERENCES

Agoramoorthy, Govindasamy. 1986. "Recent observations on twelve cases of infanticide in Hanuman langur, *Presbytis entellus,* around Jodhpur, India." *Primate Report* 14:209–10.———. 1994. "Adult male replacement and social change in two troops of Hanuman langur (*Presbytis entellus*) at Jodhpur, India." *International Journal of Primatology* 15:225–38.

Agoramoorthy, Govindasamy, and S. M. Mohnot. 1988. "Infanticide and juvenilicide in Hanuman langurs (*Presbytis entellus*) around Jodhpur, India." *Human Evolution* 3:279–96.

Agoramoorthy, Govindasamy, et al. 1988. "Abortions in free-ranging Hanuman langurs (*Presbytis entellus*)—a male induced strategy?" *Human Evolution* 3:297–308.

Agrell, Jep, Jerry O. Wolff, and Hannu Ylönen. 1998. "Counter-strategies to infanticide in mammals: Costs and consequences." *Oikos* 83:507–17.

Alcock, John. 1975. *Animal behavior: An evolutionary approach.* Sunderland, MA: Sinauer.

———. 1979. *Animal behavior: An evolutionary approach.* 2d ed. Sunderland, MA: Sinauer.

Aldrich-Blake, F. P. G. 1970. "Problems of social structure in forest monkeys." In *Social behaviour in birds and mammals,* ed. John H. Crook, 79–101. London: Academic Press.

Allen, Elizabeth, et al. 1975. "Against 'Sociobiology.'" *New York Review of Books,* 22 (18): 43–44.

Altmann, Jeanne. 1974. "Observational study of behaviour: Sampling methods." *Behaviour,* 49:227–67.

Altmann, Stuart A. 1959. "Field observations on a howling monkey society." *Journal of Mammalogy* 40:317–30.

———. 1962a. "A field study of the sociobiology of rhesus monkeys." *Annals of the New York Academy of Sciences* 12:338–435.

———.1962b. "Social behavior of anthropoid primates: Analysis of recent concepts." In *Roots of behavior: Genetics, instinct and socialization in animal behavior,* ed. Eugene L. Bliss, 277–85. New York: Harper and Brothers.

———, ed. 1967. *Social communication among primates.* Chicago: University of Chicago Press.

Andelman, S. J. 1987. "Evolution of concealed ovulation in vervet monkeys (*Cercopithecus aethiops*)." *American Naturalist* 129:785–99.

Angst, W., and D. Thommen. 1977. "New data and a discussion of infant killing in Old World monkeys and apes." *Folia Primatologica* 27:198–229.

Anon. 1963. "Preface." *Folia Primatologica* 1:1.

Anon. 1964. "The regional primate center program." *Folia Primatologica* 2:124–28.

Arcadi, A. C., and Richard Wrangham. 1999. "Infanticide in chimpanzees: Review of cases and a new within-group observation from the Kanyawara study group in Kibale National Park." *Primates* 40:337–51.

Asquith, Pamela. 1994. "The intellectual history of field studies in primatology, East and West." In *Strength in diversity: A reader in physical anthropology,* ed. L. K. Chan and A. Herring, 49–75. Toronto: Canadian Scholars' Press.

———. 1997. "Why anthropomorphism is *not* metaphor: Crossing concepts and cultures in animal behavior studies." In *Anthropomorphism, anecdotes, and animals: The emperor's new clothes? ed.* R. W. Mitchell, N. S. Thompson, and H. Lyn Miles, 22–34. New York: SUNY.

———. 2000. "Negotiating science: Internationalization and Japanese primatology." In *Primate encounters: Models of science, gender, and society,* ed. Shirley Strum and Linda Fedigan, 165–83. Chicago: University of Chicago Press.

Baldwin, J. D., and J. I. Baldwin. 1972a. "The ecology and behavior of squirrel monkeys (*Saimiri oerstedi*) in a natural forest in western Panama." *Folia Primatologica* 18:161–84.

———. 1972b. "Population density and use of space in howling monkeys (*Alouatta villosa*) in south west Panama." *Primates* 13:371–79.

———. 1973. "Interactions between adult female and infant howling monkeys (*Alouatta palliata*)." *Folia Primatologica* 20:27–72.

Barnes, Barry. 1982. *T. S. Kuhn and social science.* London: Macmillan.

———. 1990. "Thomas Kuhn." In *The return of grand theory in the human sciences,* ed. Quentin Skinner, 83–100. Cambridge: Cambridge University Press.

Barnes, Barry, and David Bloor. 1982. "Relativism, rationalism and the sociology of knowledge." In *Rationality and relativism,* ed. M. Hollis and S. Lukes, 21–47. Oxford: Blackwell.

Barnes, Barry, David Bloor, and John Henry. 1996. *Scientific knowledge: A sociological analysis.* Chicago: University of Chicago Press.

Bartlett, Thad, Robert Sussman, and James Cheverud. 1993. "Infant killing in primates: A review of observed cases with specific reference to the sexual selection hypothesis." *American Anthropologist* 95:958–90.

Bates, B. C. 1970. "Territorial behaviour in primates: A review of recent field studies." *Primates* 11:271–84.

Beatty, H 1951. "A note on the behavior of the chimpanzee." *Journal of Mammalogy* 32:118.

Bernstein, Irwin. 1967. "Intertaxa interactions in a Malaysian primate community." *Folia Primatologica* 7:198–207.

———. 1968. "The lutong of Kuala Selangor." *Behaviour* 32:1–15.

Bertram, Brian. 1975. "Social factors influencing reproduction in wild lions." *Journal of Zoology* 177:463–82.

———. 1976. "Kin selection in lions and in evolution." In *Growing points in ethology,* ed. P. Bateson and R. Hinde, 281–301. Cambridge: Cambridge University Press.

Bird, Alexander. 2000. *Thomas Kuhn.* Princeton, NJ: Princeton University Press.

Bishop, Naomi. 1979. "Himalayan langurs: Temperate colobines." *Journal of Human Evolution* 8:251–81.

Bishop, Naomi, Sarah Hrdy, J. Teas, and Jim Moore. 1981. "Measures of human influence in habitats of South Asian monkeys." *International Journal of Primatology* 2:153–67.

Bloor, David. 1991/1976. *Knowledge and social imagery.* Chicago: University of Chicago Press.

Boer, Michael, and Volker Sommer. 1992. "Evidence for sexually selected infanticide in captive *Cercopithecus mitis, Cercocebus torquatus,* and *Mandrillus leucophaeus.*" *Primates* 33:557–63.

Boesch, Christophe, and Hedwige Boesch-Achermann. 2000. *The chimpanzees of the Taï Forest: Behavioural ecology and evolution,* Oxford: Oxford University Press.

Boggess, Jane. 1979. "Troop male membership and infant killing in langurs (*Presbytis entellus*)." *Folia Primatologica* 32:65–107.

———. 1984. "Infant killing and male reproductive strategies in langurs." In *Infanticide: Comparative and evolutionary perspectives,* ed. Glenn Hausfater and Sarah Hrdy, 283–310. New York: Aldine.

Bolwig, N. 1959a. "A study of the nests built by mountain gorilla and chimpanzee." *South African Journal of Science* 55 (11): 286–91.

———. 1959b. "A study of the behaviour of the chacma baboon, *Papio ursinus.*" *Behaviour* 14:136–63.

Booth, C. 1962. "Some observations on the behavior of *Cercopithecus* monkeys." *Annals of the New York Academy of Sciences* 102:477–87.

Borries, Carola. 1988. "Patterns of grand-maternal behaviour in free-ranging Hanuman langur (*Presbytis entellus*)." *Human Evolution* 3:239–60.

———. 1997. "Infanticide in seasonally breeding multimale groups of Hanuman langurs (*Presbytis entellus*) in Ramnagar (South Nepal)." *Behavioral Ecology and Sociobiology* 41:139–50.

Borries, Carola, et al. 1999a. "Males as infant protectors in Hanuman langurs (*Presbytis entellus*) living in multimale groups: Defence pattern, paternity and sexual behaviour." *Behavioral Ecology and Sociobiology* 46:350–56.

———. 1999b. "DNA analyses support the hypothesis that infanticide is adaptive in langur monkeys." *Proceedings of the Royal Society of London,* ser. B, 266:901–4.

Bourlière, F. 1962. "Patterns of social grouping among wild primates." In *Social life of early man,* ed. S. L. Washburn, 1–10. London: Methuen.

———. 1968. "Lemur behavior." *American Anthropologist* 70:648–49.

Breden, F., and Glen Hausfater. 1990. "Selection within and between groups for infanticide." *American Naturalist* 136:673–88.

Brotherton, Peter N. M., and Anna Rhodes. 1996. "Monogamy without biparental care in a dwarf antelope." *Proceedings of the Royal Society of London,* ser. B, 263:23–29.

Browne, Janet. 1996. "Biogeography and empire." In *Cultures of natural history,* ed. N. Jardine, J. A. Secord, and E. Spary, 305–21. Cambridge: Cambridge University Press,

Burkhardt, Richard. 1999. "Ethology, natural history, the life sciences, and the problem of place." *Journal of the History of Biology* 32:489–508.

———. 2005. *Patterns of behavior: Konrad Lorenz, Niko Tinbergen and the founding of ethology.* Chicago: University of Chicago Press.

Burt, Jonathan. 2003. *Animals in film.* London: Reaktion Books.

———. 2006. "Solly Zuckerman: The making of a primatological career in Britain, 1925–1945." *Studies in the History and Philosophy of the Biological and Biomedical Sciences* 37 (2): 295–310.

Busse, C. D. 1985. "Paternity recognition in multi-male primate groups." *American Zoologist* 25:873–81.

Busse, C. [D.], and W. Hamilton. 1981. "Infant carrying by male chacma baboons." *Science* 212:1281–83.

Butler, H. 1966a. "Observations on the menstrual cycle of the grivet monkey (*Cercopithecus aethiops aethiops*) in the Sudan." *Folia Primatologica* 4:194–205.

———. 1966b. "Some notes on the distribution of primates in the Sudan." *Folia Primatologica* 4:416–23.

Butler, H. 1967. "Seasonal breeding of the Senegal galago (*Galago senegalensis senegalensis*) in the Nubia Mountains, Republic of the Sudan." *Folia Primatologica* 5:165–75.

Butynski, Thomas. M. 1982. "Harem male replacement and infanticide in the blue monkey (*Cercopithecus mitis stuhlmanni*) in the Kibale Forest, Uganda." *American Journal of Primatology* 3:1–22.

———. 1990. "Comparative ecology of blue monkeys (*Cercopithecus mitis*) in high and low density subpopulations." *Ecological Monographs,* 1–26.

Bygott, David. 1972. "Cannibalism among wild chimpanzees." *Nature* 238:410–11.

Calhoun, John. 1962a. "Population density and social pathology." *Scientific American* 206 (2): 139–48.

———. 1962b. "A behavioral sink." In *Roots of behavior: Genetics, instinct, and socialization in animal behavior,* ed. Eugene L. Bliss, 295–325. New York: Harper Brothers.

Camerini, J. 1996. "Wallace in the Field." *Osiris* 11:44–65.

Campbell, Bob. 2000. *The taming of the gorillas.* London: Minerva Press.

Cannon, Susan Faye. 1978. *Science in culture: The early Victorian period.* New York: Science History Publications.

Carey, D. 1997. "Compiling nature's history: Travellers and travel narratives in the early Royal Society." *Annals of Science* 54:269–92.

Carpenter, Clarence Ray. 1962. "Comments, problems, suggestions." *Annals of the New York Academy of Sciences* 102:488–96.

———. 1964. *Naturalistic behavior of nonhuman primates.* University Park: Pennsylvania State University Press.

———. 1965. "The howlers of Barro Colorado Island." In *Primate behavior: Field studies of monkeys and apes,* ed. Irven DeVore, 250–91. New York: Holt, Rinehart and Winston.

———, ed. 1969. *Proceedings of the Second International Congress of Primatology.* Basel: Karger.

Chalmers, N. R. 1968a. "Group composition, ecology and daily activities of free-living mangabeys in Uganda." *Folia Primatologica* 8:247–62.

———. 1968b. "The social behavior of free living mangabeys in Uganda." *Folia Primatologica* 8:263–81.

———. 1986. "Book review: *Infanticide: Comparative and evolutionary perspectives.*" *International Journal of Primatology* 7:327–29.

Chance, M. R. 1955. "The sociability of monkeys." *Man* 55:162–65.

———. 1956. "Social structure of a colony of *Macaca mulatta.*" *British Journal of Animal Behaviour* 4:1–13.

———. 1962. "The nature and special features of the instinctive social bond of primates." In *Social life of early man,* ed. S. L. Washburn, 17–33. London: Methuen.

Chance, M. R., and A. P. Mead. 1953. "Social behaviour and primate evolution." *Symposium of the Society of Experimental Biology* 7:395–439.

Chapman, M., and Glenn Hausfater. 1979. "The reproductive consequences of infanticide in langurs: A mathematical model." *Behavioral Ecology and Sociobiology* 5:227–40.

Cheney, Dorothy, and Robert Seyfarth. 1990. *How monkeys see the world.* Chicago: University of Chicago Press.

Cheney, Dorothy, and Richard Wrangham. 1987. "Predation." In *Primate societies,* ed. Barbara Smuts et al., 227–39. Chicago: University of Chicago Press.

Cheney, Dorothy, et al. 1987. "The study of primate societies." In *Primate societies,* ed. Barbara Smuts et al., 1–8. Chicago: University of Chicago Press.

Ciani, A. 1984. "A case of infanticide in a free-ranging group of rhesus monkeys (*Macaca mulatta*) in the Jackaroo Forest, Simla, India." *Primates* 25:372–77.

Clarke, M. 1983. "Infant killing and infant disappearance following male takeovers in a group of free ranging howling monkeys (*Alouatta palliata*) in Costa Rica." *American Journal of Primatology* 5:241–57.

Clutton-Brock, T., P. N. M. Brotherton, R. Smith, G. M. McIlrath, R. Kansky, D. Gaynor, M. J. O'Riain, and J. D. Skinner. 1998. "Infanticide and expulsion of females in a co-operative mammal." *Proceedings of the Royal Society of London,* ser. B, 265:2291–95.

Clutton-Brock, T., and G. A. Parker. 1995. "Sexual coercion in animal societies." *Animal Behaviour* 49:1345–65.

Cohen, I. B. 1985. *Revolutions in science.* Cambridge, MA: Harvard University Press.

Cole, Simon. 1996. "Which came first, the fossil or the fuel?" *Social Studies of Science* 26:733–66.

Collias, N. 1951. "Problems and principles of animal sociology." In *Comparative psychology,* ed. C. P. Stone, 389–421. New York: Prentice-Hall.

Collias, N., and C. Southwick. 1952. "A field study of population density and social organisation in howling monkeys." *Proceedings of the American Philosophical Society* 96 (2): 143–56.

Collins, Harry. 1992/1985. *Changing order: Replication and induction in scientific practice.* Chicago: University of Chicago Press.

———. 1999. "Tantalus and the aliens: Publications, audiences and the search for gravitational waves." *Social Studies of Science* 29:163–97.

———. 2004. *Gravity's shadow: The search for gravitational waves.* Chicago: University of Chicago Press.

Collins, Harry, and Robert Evans. 2002. "The third wave of science studies: Studies of expertise and experience." *Social Studies of Science* 32:235–96.

Collins, Harry, and Trevor Pinch. 1982. *Frames of meaning: The social construction of extraordinary science.* London: Routledge and Kegan Paul.

———. 1993. *The golem: What everyone should know about science.* Cambridge: Cambridge University Press.

Conaway, Clinton, and Carl Koford. 1964. "Estrous cycles and mating behavior in a free-ranging band of rhesus monkeys." *Journal of Mammalogy* 45:577–88.

Crist, Eileen, 1996. "Naturalists' portrayals of animal life: Engaging the verstehen approach." *Social Studies of Science* 26:799–838.

———. 1999. *Images of animals: Anthropomorphism and animal mind.* Philadelphia: Temple University Press.

Crook, John. 1970. *Social behaviour in birds and mammals.* London: Academic Press.

Crook, John, and Stephen Gartlan. 1966. "Evolution of primate societies." *Nature* 210:1200–1203.

Curtin, Richard. 1977. "Langur social behavior and infant mortality." *Kroeber Anthropological Papers* 50:27–36.

———. 1981. "Strategy and tactics in grey male langur competition." *Journal of Human Evolution* 10:245–53.

———. 1982. "Females, male competition and gray langur troop structure." *Folia Primatologica* 37:216–27.

Curtin, Richard, and Phyllis Dolhinow. 1978. "Primate social behavior in a changing world." *American Scientist* 66:468–75.

Dagg, Anne. 1998. "Infanticide by male lions hypothesis: A fallacy influencing research into human behavior." *American Anthropologist* 100:940–50.

———. 1999. "Sexual selection is debatable." *Anthropology News,* December, 20.

———. 2000. "The infanticide hypothesis: A response to the response." *American Anthropologist* 102:831–34.

Daly, M., and M. Wilson. 1988. "Evolutionary social psychology and family homicide." *Science* 242:519–24.

Dart, R. A. 1960. "Can the mountain gorilla be saved?" *Current Anthropology* 1:330–32.

———. 1961. "The Kisoro pattern of mountain gorilla preservation." *Current Anthropology* 2:510–11.

Daston, Lorraine, and Greg Mitman, eds. 2006. *Thinking with animals: New perspectives on anthropomorphism.* New York: Columbia University Press.

Davenport, R. K. 1967. "The orang-utan in Sabah." *Folia Primatologica* 5:247–63.

Davies, A. G. 1987. "Adult male replacement and group formation in *Presbytis rubicunda.*" *Folia Primatologica* 49:111–14.

Dettlebach, M. 1996. "Humboldtian science." In *Cultures of natural history,* ed. N. Jardine, J. A. Secord, and E. Spary, 287–304. Cambridge: Cambridge University Press.

DeVore, Irven. 1963. "Mother-infant relationships in free-ranging baboons." In *Maternal behavior in mammals,* ed. H. L. Rheingold, 305–35. New York: John Wiley.

———, ed. 1965. *Primate behavior: Field studies of monkeys and apes.* New York: Holt, Rinehart and Winston.

DeVore, Irven, and K. R. L. Hall. 1965. "Baboon ecology." In *Primate behavior: Field studies of monkeys and apes,* ed. Irven DeVore, 20–52. New York: Holt, Rinehart and Winston.

DeVore, Irven, and Richard Lee. 1963. "Recent and current field studies of primates." *Folia Primatologica* 1:66–72.

DeVore, Irven, and Sherwood Washburn. 1964. "Baboon ecology and human evolution." In *African ecology and human evolution,* ed. F. C. Howell, 335–67. London: Methuen.

Dickeman, Mildred. 1975. "Demographic consequences of infanticide in man." *Annual Review of Ecology and Systematics* 6:107–37.

Digby, Leslie. 1995. "Infant care, infanticide and female reproductive strategies in polygynous groups of common marmosets (*Callithrix jacchus*)." *Behavioral Ecology and Sociobiology* 37:51–61.

———. 1999. "Sexual behavior and extra-group copulations in a wild population of common marmosets (*Callithrix jacchus*)." *Folia Primatologica* 70:136–45.

Dolhinow, Phyllis, ed. 1972. *Primate patterns.* New York: Holt, Rinehart and Winston.

———. 1977. "Normal monkeys?" *American Scientist* 65:266.

———. 1980. "An experimental study of mother loss in the Indian langur monkey (*Presbytis entellus*)." *Folia Primatologica* 33:77–128.

———. 1999. "Understanding behavior: A langur monkey case study." In *The nonhuman primates,* ed. Phyllis Dolhinow and Augustain Fuentes, 189–95. Mountain View, CA: Mayfield.

Dolhinow, Phyllis, and Mark A. Taff. 1993. "Immature and adult langur monkey (*Presbytis entellus*) males: Infant-initiated adoption in a colony group." *International Journal of Primatology* 14:919–26.

Donisthorpe, J. H. 1958. "A pilot study of the mountain gorilla (*Gorilla gorilla beringei*) in South West Uganda, February to September, 1957." *South African Journal of Science* 54:195–217.

Douglas-Hamilton, Iain, and Oria Douglas-Hamilton. 1975. *Among the elephants.* London: Collins.

Dukelow, W. Richard. 1995. *The alpha males: An early history of the regional primate research centres.* Lanham, MD: University Press of America.

Dunbar, R. I. M. 1988. *Primate social systems.* London: Croom Helm.

Duncan, K. G. 1987. "The zoological exploration of the Australian region and its impact on biological theory." In *Scientific colonialism: A cross-cultural comparison,* ed. N. Reingold and M. Rothenberg, 79–100. Washington, DC: Smithsonian Institution Press.

Eisenberg, J. F., N. A. Muckenhirn, and R. Rudran. 1972. "The relation between ecology and social structure in primates." *Science* 176:863–74.

Emlen, J. T., and George Schaller. 1960. "Current field studies of the mountain gorilla." *South African Journal of Science* 56 (4): 88–89.

———. 1963. "In the home of the mountain gorilla." In *Primate social behavior,* ed. Charles Southwick, 124–35. Princeton, NJ: Van Nostrand.

Fairgrieve, Chris. 1995. "Infanticide and infant eating in the blue monkey (*Cercopithecus mitis stuhlmanni*) in the Budongo Forest Reserve, Uganda." *Folia Primatologica* 64:69–72.

Fedigan, Linda. 1992/1982. *Primate paradigms: Sex roles and social bonds.* Chicago: University of Chicago Press. Originally published Montreal: Eden Press.

Findlen, Paula. 1994. *Possessing nature: Museums, collecting and scientific cultures in early modern Italy.* Berkeley: University of California Press.

Fleck, Ludwig. 1979/1935. *Genesis and development of a scientific fact.* Chicago: University of Chicago Press.

Fossey, Dian. 1983. *Gorillas in the mist.* London: Hodder and Stoughton.

Franklin, Adrian. 1999. *Animals and modern cultures: A sociology of human-animal relations in modernity,* London: Sage.

Franklin, Allen. 1990. *Experiment, right or wrong.* Cambridge: Cambridge University Press.

Freed, Leonard A. 1987. "Prospective infanticide and protection of genetic paternity in tropical house wrens." *American Naturalist* 130 (6): 948–54.

Frisch, J. E. 1959. "Research on primate behavior in Japan." *American Anthropologist* 61:584–96.

Fuller, Steve. 1997. *Science.* Buckingham: Open University Press.

———. 2000. *Thomas Kuhn: A philosophical history for our times.* Chicago: University of Chicago Press.

———. 2006. *The new sociological imagination.* London: Sage.

Gartlan, J. S. 1968. "Structure and function in primate society." *Folia Primatologica* 8:89–120.

Ghiglieri, Michael. 1988. *East of the Mountains of the Moon: Chimpanzee society in the African rainforest.* New York: Free Press.

Gieryn, Thomas. 1992. "The ballad of Pons and Fleischmann: Experiment and narrativity in the (un)making of cold fusion." In *The social dimensions of science,* ed. E. McMullin, 217–43. Nortre Dame, IN: University of Notre Dame Press.

———. 1999. *Cultural boundaries of science: Credibility on the line.* Chicago: University of Chicago Press.

———. 2006. "City as truth-spot: Laboratories and field sites in urban studies." *Social Studies of Science* 36:5–38.

Gilbert D., C. Packer, A. Pusey, and S. O'Brian. 1991. "Analytical DNA fingerprinting in lions: Parentage, genetic diversity and kinship." *Journal of Heredity* 82:378–86.

Gilbert, Nigel, and Michael Mulkay. 1984. *Opening Pandora's box: A sociological analysis of scientists' discourse.* Cambridge: Cambridge University Press.

Glass, Gregory, Robert Hold, and Norman Slade. 1995. "Infanticide as an evolutionary stable strategy." *Animal Behaviour* 33:384–91.

Goldschmidt, Walter. 1965. "Book review: *The mountain gorilla.*" *Current Anthropology* 6:297–98.

Golinski, Jan. 1998. *Making natural knowledge: Constructivism and the history of science.* Cambridge: Cambridge University Press.

Gomendio, Montserrat, and Fernando Colmenares. 1989. "Infant killing and infant adoption following the introduction of new males to an all-female colony of baboons." *Ethology* 80:223–44.

Goodall, Jane. 1962. "Nest-building behavior in the free-ranging chimpanzee." *Annals of the New York Academy of Sciences* 102:455–67.

———. 1965. "Chimpanzees of the Gombe Stream Reserve." In *Primate behavior: Field studies of apes and monkeys,* ed. Irven DeVore, 425–73. New York: Holt, Rinehart and Winston.

———. 1971. *In the shadow of man.* Boston: Houghton Mifflin.

———. 1977. "Infant killing and cannibalism in free-living chimpanzees." *Folia Primatologica* 28:259–82.

———. 1979. "Life and death at Gombe." *National Geographic* 155 (May): 592–621.

———. 1986. *The chimpanzees of Gombe.* Cambridge, MA: Belknap Press.

———. 1990. *Through a window: Thirty years with the chimpanzees of Gombe.* Boston: Houghton Mifflin.

Goodall, Jane, and Dale Peterson. 1993. *Visions of Caliban: Of chimpanzees and people.* Athens: University of Georgia Press.

Grafen, A. 1982. "How not to measure inclusive fitness." *Nature* 298:425.

Grimshaw A., and K. Hart. 1991. *Anthropology and the crisis of the intellectuals.* Cambridge: Prickly Pear Press.

Gross, Paul, and Norman Levitt. 1994. *Higher superstition: The academic left and its quarrels with science.* Baltimore: Johns Hopkins University Press.

Gross, Paul, Norman Levitt, and Martin Lewis. 1997. *The flight from science and reason.* New York: New York Academy of Sciences.

Gupta, Akhil, and James Ferguson, eds. 1997. *Anthropological locations: Boundaries and grounds of a field science.* Berkeley: University of California Press.

Haddow, A. J. 1952. "Field and laboratory studies on an African monkey." *Proceedings of the Zoological Society of London* 122:297–394.

Hall, K. R. L. 1960. "Social vigilance behaviour of the chacma baboon, *Papio ursinus.*" *Behaviour* 16:261–94.

———. 1962. "Numerical data, maintenance activities and locomotion of the wild chacma baboon." *Proceedings of the Zoological Society of London* 139:181–220.

———. 1963. "Variations in the ecology of the chacma baboon, *Papio ursinus.*" *Symposia of the Zoological Society of London* 10:1–28.

———. 1968a. "Social organization of the Old World monkeys and apes." In *Primates: Studies in adaptation and variability,* ed. Phyllis Jay, 7–31. New York: Holt, Rinehart and Winston,

———. 1968b. "Experiment and quantification in the study of baboon behavior in its natural habitat." In *Primates: Studies in adaptation and variability,* ed. Phyllis Jay, 120–30. New York: Holt, Rinehart and Winston.

———. 1968c. "Behavior and ecology of the wild patas monkey, *Erythrocebus patas,* in Uganda." In *Primates: Studies in adaptation and variability,* ed. Phyllis Jay, 32–119. New York: Holt, Rinehart and Winston.

———. 1968d. "Social learning in monkeys." In *Primates: Studies in adaptation and variability,* ed. Phyllis Jay, 383–97. New York: Holt, Rinehart and Winston.

Hall, K. R. L., and Irven DeVore. 1965. "Baboon social behavior." In *Primate behavior: Field studies of monkeys and apes,* ed. Irven DeVore, 53–110. New York: Holt, Rinehart and Winston.

Ham, J., and M. Senior. 1997. *Animal acts: Configuring the human in Western history.* New York: Routledge.

Hamai, Miya, et al. 1992. "New records of within-group infanticide and cannibalism in wild chimpanzees." *Primates* 33:151–62.

Hamilton, W. D. 1964a. "The genetical evolution of social behaviour I." *Journal of Theoretical Biology* 7:1–16.

———. 1964b. The genetical evolution of social behaviour II." *Journal of Theoretical Biology* 7:17–32.

Hanby, J. P., and Bygott, J. D. 1979. "Population changes in lions and other predators." In *Serengeti: Dynamics of an ecosystem,* ed. A. E. R. Sinclair and M. Norton-Griffiths, 249–62. Chicago: University of Chicago Press.

———. 1987. "Emigration of subadult lions." *Animal Behaviour* 35:161–69.

Hannaway, O. 1986. "Laboratory design and the aim of science: Andreas Libavius versus Tycho Brahe." *Isis* 77:585–610.

Haraway, Donna. 1983. "The contest for primate nature: Daughters of man the hunter in the field, 1960–80." In *The future of American democracy: Views from the left,* ed. Mark Kann, 175–207. Philadelphia: Temple University Press.

———. 1990. *Primate visions.* London: Verso.

———. 2003. *The companion species manifesto: Dogs, people and significant others.* Chicago: Prickly Paradigm Press.

Harcourt, A. H., and J. Greenberg. 2001. "Do gorilla females join males to avoid infanticide? A quantitative model." *Animal Behaviour* 62:905–15.

Hausfater, Glenn. 1977. "Primate infanticide." *American Scientist* 65:404.

———. 1984. "Infanticide in nonhuman primates: An introduction and perspective." In *Infanticide: Comparative and evolutionary perspectives,* ed. Glenn Hausfater and Sarah Hrdy, 145–50. New York: Aldine.

Hausfater, Glenn, and Sarah Hrdy, eds. 1984. *Infanticide: Comparative and evolutionary perspectives.* New York: Aldine.

Hausfater, Glenn, and Christian Vogel. 1982. "Infanticide in langur monkeys (genus *Presbytis*): Recent research and a review of hypotheses." In *Advanced views in primate biology,* ed. A. B. Chiarelli and R. S. Corruccini, 160–76. Berlin: Springer Verlag.

Henke, Christopher. 2000. "Making a place for science: The field trial." *Social Studies of Science* 30:483–511.

Henninger-Voss, M., ed. 2002. *Animals in human histories: The mirror of nature and culture.* Rochester, NY: University of Rochester Press.

Henzi, S. P. 1988. "Many males do not a multi-male troop make." *Folia Primatologica* 51:165–68.

Hevly, Bruce. 1996. "The heroic science of glacier motion." *Osiris* 11:66–86.

Hinde, Robert. 2000. "Some reflections on primatology at Cambridge and the science studies debate." In *Primate encounters: Models of science, gender, and society,* ed. Shirley Strum and Linda Fedigan, 104–15. Chicago: University of Chicago Press.

Hiraiwa-Hasegawa, Mariko. 1987. "Infanticide in primates and a possible case of male-biased infanticide in chimpanzees." In *Animal societies: Theories and facts,* ed. Y. Ito et al., 125–39. Tokyo: Japan Science Society Press.

Hiraiwa-Hasegawa, Mariko, and T. Hasegawa. 1994. "Infanticide in nonhuman primates: Sexual selection and local resource competition." In *Infanticide and parental care,* ed. S. Parmigiani and F. S. vom Saal, 137–54. New York: Harwood.

Holloway, Ralph L., ed. 1974. *Primate aggression, territoriality and xenophobia: A comparative perspective.* New York: Academic Press.

Hoogland, John L. 1985. "Infanticide in prairie dogs: Lactating females kill offspring of close kin." *Science* 230:1037–40.

Hooton, Ernest A. 1954. "The importance of primate studies in anthropology." *Human biology* 26 (3): 179–88.

Hrdy, Sarah. 1974. "Male-male competition and infanticide among the langurs (*Presbytis entellus*) of Abu, Rajasthan." *Folia Primatologica* 22:19–58.

———.1977a. *The langurs of Abu: Female and male strategies of reproduction.* Cambridge, MA: Harvard University Press.

———. 1977b. "Infanticide as a primate reproductive strategy." *American Scientist* 65:40–49.

———. 1978. "More monkey business." *American Scientist* 66:667–68.

———. 1979. "Infanticide among animals: A review, classification, and examination of the implications for the reproductive strategies of females." *Ethology and Sociobiology* 1:13–40.

———. 1982. "Positivist thinking encounters field primatology, resulting in agonistic behaviour." *Social Science Information* 21:245–50.

———. 1984. "Assumptions and evidence regarding the sexual selection hypothesis: A reply to Boggess." In *Infanticide: Comparative and evolutionary perspectives,* ed. Glenn Hausfater and Sarah Hrdy, 315–19. New York: Aldine.

———. 1997. "Raising Darwin's consciousness: Female sexuality and the prehominid origins of patriarchy." *Human Nature* 8:1–49.

———. 2000. "Foreword." In *Infanticide by males and its implications,* ed. Carel van Schaik and Charles Janson, xi–xiv. Cambridge: Cambridge University Press.

Hrdy, Sarah, and Glenn Hausfater. 1984. "Comparative and evolutionary perspectives on infanticide: Introduction and overview." In *Infanticide: Comparative and evolutionary perspectives,* ed. Glenn Hausfater and Sarah Hrdy, xiii–xxxv. New York: Aldine.

Hrdy, Sarah, Charles Janson, and Carel van Schaik. 1995. "Infanticide: Let's not throw out the baby with the bathwater." *Evolutionary Anthropology* 3:151–54.

Hughes, T. H. 1884. "An incident in the habits of the *Semnopithecus entellus,* the common Hanuman monkey." *Proceedings of the Asiatic Society of Bengal,* 147–50.

Imanishi, K. 1960. "Social organization of nonhuman primates in their natural habitat." *Current Anthropology* 1:393–407.

———. 1963/1957. "Social behavior of Japanese monkeys." In *Primate social behavior,* ed. Charles Southwick, 68–81. Princeton, NJ: Van Nostrand, Originally published in *Psychologica.*

Itani, J. 1963. "Paternal care in the wild Japanese monkey, *Macaca fuscata.*" In *Primate social behavior,* ed. Charles Southwick, 91–97. Princeton, NJ: Van Nostrand.

———. 1972. "A preliminary essay on the relationships between social organization and incest avoidance in non-human primates." In *Primate socialization,* ed. F Poirer, 165–71. New York: Random House.

Jackson, Stevi, and Amanda Rees. 2007. "The appalling appeal of nature: The popular influence of evolutionary psychology as a problem for sociology." *Sociology* 41:917–30.

Janson, Charles. 2000. "Primate socioecology: The end of a golden age?" *Evolutionary Anthropology* 9:73–86.

Jasanoff, Sheila. 1996. "Beyond epistemology: Relativism and engagement in the politics of science. *Social Studies of Science* 26:393–418.

Jay, Phyllis. 1962. "Aspects of maternal behavior among langurs." *Annals of the New York Academy of Sciences* 102:468–76.

———. 1963a. "The social behavior of the langur monkey." PhD diss., University of Chicago.

———. 1963b. "Mother-infant relations in langurs." In *Maternal behavior in mammals,* ed. H. L. Rheingold, 282–304. New York: Wiley.

———. 1963c. "The Indian langur monkey." In *Primate social behavior,* ed. Charles Southwick, 114–23. Princeton, NJ: Van Nostrand.

———. 1965. "The common langur of North India." In *Primate behavior: Field studies of monkeys and apes,* ed. Irven DeVore, 197–249, New York: Holt, Rinehart and Winston.

———, ed. 1968. *Primates: Studies in adaptation and variability.* New York: Holt, Rinehart and Winston.

J. F. G. 1901. "Habits of the Lungoor monkey." *Journal of the Bombay Natural History Society* 14:149–51.

Jolly, Alison. 1966a. "Lemur social behavior and primate intelligence." *Science* 153:591–606.

———. 1966b. *Lemur behavior: A Madagascar field study.* Chicago: University of Chicago Press.

———. 1972. *The evolution of primate behavior.* New York: Macmillan.

———. 2000. "The bad old days of primatology?" In *Primate encounters: Models of science, gender, and society,* ed. Shirley Strum and Linda Fedigan, 71–84. Chicago: University of Chicago Press.

Jolly, Alison, et al. 2000. "Infant killing, wounding and predation in *Eulemur* and *Lemur.*" *International Journal of Primatology* 21:20–40.

Kappeler, P. 2002, "Sexual selection in primates: New and comparative perspectives." *Evolutionary Anthropology* 11:173–75.

Kawabe, M, 1970. "A preliminary study of the wild siamang gibbon (*Hylobates syndactylus*) at Fraser's Hill, Malaysia." *Primates* 11:285–91.

Kawabe, M., and T. Mano. 1972. "Ecology and behaviour of the wild proboscis monkey, *Nasalius larvatus,* (Wurmb) in Sahab, Malaysia." *Primates* 13:213–28.

Kawanaka, K. 1981. "Infanticide and cannibalism in chimpanzees, with special reference to the newly observed case in the Mahale Mountains." *African Study Monographs* 1:69–91.

Knorr Cetina, Karin. 1983. "The ethnographic study of scientific work: Towards a constructivist interpretation of science." In *Science observed: Perspectives on the social study of science,* ed. Karin Knorr Cetina and M. Mulkay, 115–40. London: Sage.

Kohler, Robert. 2002a. "Place and practice in field biology." *History of Science* 40:189–210.

———. 2002b. *Landscapes and labscapes: Exploring the lab-field border in biology.* Chicago: University of Chicago Press.

Kohler, Wolfgang. 1925. *The mentality of apes.* New York: Harcourt Brace.

Kortlandt, Adriaan. 1962. "Chimpanzees in the wild." *Scientific American* 206:128–38.

Koyama, N. 1971. "Observations on mating behavior of wild siamang gibbons at Fraser's Hill, Malaysia." *Primates* 12:183–89.

Kuhn, Thomas S. 1962. *The structure of scientific revolutions.* Chicago: University of Chicago Press.

Kuklick, Henrika. 1991. *The savage within: A social history of British anthropology, 1885–1945.* Cambridge: Cambridge University Press.

———. 1997. "After Ishmael: The fieldwork tradition and its future." In *Anthropological locations: Boundaries and grounds of a field science,* ed. Akhil Gupta and James Ferguson, 47–65. Berkeley: University of California Press.

Kuklick, Henrika, and Robert Kohler. 1996. "Introduction." *Osiris* 11:1–14.

Kummer, Hans. 1968. "Two variations in the social organization of baboons." In *Primates: Studies in adaptation and variability,* ed. Phyllis Jay, 293–321. New York: Holt, Rinehart and Winston.

———. 1995. *In quest of the sacred baboon: A scientist's journey.* Trans. M. Ann Biederman-Thorson. Princeton NJ: Princeton University Press.

Kummer, Hans, and F. Kurt. 1963. "Social units of a free-living population of hamadryas baboons." *Folia Primatologica* 1:4–19.

Labinger, Jay, and Harry Collins, eds. 2001. *The one culture? A conversation about science.* Chicago: University of Chicago Press.

Labov, Jay, William Huck, Robert Elwood, and Ronald Brooks. 1985. "Current problems in the study of infanticidal behavior in rodents." *Quarterly Review of Biology* 60:1–20.

Lancaster, Jane, and Richard Lee. 1965. "The annual reproductive cycle in monkeys and apes." In *Primate behavior: Field studies of monkeys and apes,* ed. Irven DeVore, 486–513. New York: Holt, Rinehart and Winston.

Landau, Misa. 1984. "Human evolution as narrative." *American Scientist* 72:262–68.

———. 1991. *Narratives of human evolution.* New Haven, CT: Yale University Press.

Latour, Bruno. 1987. *Science in action: How to follow scientists and engineers through society.* Cambridge MA: Harvard University Press.

———. 1988. *The Pasteurization of France.* Trans. A. Sheridon and J. Law. Cambridge MA: Harvard University Press.

———. 1999. *Pandora's hope: Essays on the reality of science studies.* Cambridge, MA: Harvard University Press.

Latour, Bruno, and Shirley Strum. 1986. "Human social origins: Please tell us another story." *Journal of Biological and Social Structures* 9:169–87.

Latour, Bruno, and Steve Woolgar. 1986/1979. *Laboratory life: The construction of scientific facts.* Princeton, NJ: Princeton University Press.

LeBoeuf, B. J., and C. Campagna. 1994. "Protection and abuse of young in pinnipeds." In *Infanticide and parental care,* ed. S. Parmigiani and F. S. vom Saal, 257–76. New York: Harwood.

Lewenstein, B. 1995. "From fax to facts: Communication and the cold fusion saga." *Social Studies of Science* 25:403–36.

Lorenz, Konrad. 1963. *On aggression.* London: Methuen.

Lynch, Michael. 1985. *Art and artifact in laboratory science: A study of shop work and shop talk in a research laboratory.* London: Routledge.

MacKenzie, Donald. 1981. *Statistics in Britain, 1865–1930: The social construction of scientific knowledge.* Edinburgh: University of Edinburgh Press.

———. 1990. *Inventing accuracy: A historical sociology of nuclear missile guidance,* Cambridge, MA: MIT Press.

Makwana, S. 1979. "Infanticide and social change in two groups of the Hanuman langur, *Presbytis entellus,* at Jodhpur." *Primates* 20:293–300.

Manson, J. H., J. Gros-Louis, and S. Perry. 2004. "Three apparent cases of infanticide by males in wild white-faced capuchins (*Cebus capucinus*)." *Folia Primatologica* 75:104–6.

March, E. W. 1957. "Gorillas of eastern Nigeria." *Oryx* 4 (1): 30–34.

Marks, Jonathan. 2000."Sherwood Washburn, 1911–2000." *Evolutionary Anthropology* 9:225–26.

Marler, Peter. 1965."Communication in monkeys and apes." In *Primate behavior: Field studies of monkeys and apes,* ed. Irven DeVore, 544–84. New York: Holt, Rinehart and Winston.

Martin, Brian. 1991. *Scientific knowledge in controversy: The social dynamics of the fluoridation debate.* Albany: SUNY Press.

Mason, William. 1968. "Naturalistic and experimental investigations of the social behavior of monkeys and apes." In *Primates: Studies in adaptation and variability,* ed. Phyllis Jay, 398–419. New York: Holt, Rinehart and Winston.

Maynard Smith, John. 1976. "Evolution and the theory of games." *American Scientist* 64:41–45.

Mayr, Ernst, and William B. Provine. 1980. *The evolutionary synthesis: Perspectives on the unification of biology,* Cambridge MA: Harvard University Press.

McComb, Karen, et al. 1993. "Female lions can identify potentially infanticidal males from their roars." *Proceedings of the Royal Society of London,* ser. B, 252:59–64.

McCook, S. 1996. "'It may be truth, but it is not evidence': Paul du Chaillu and the legitimation of evidence in the field sciences." *Osiris* 11:177–97.

Mestel, Rosie. 1995. "Monkey 'murderers' may be falsely accused." *New Scientist,* July 15, 17.

Mitchell, R., N. Thompson, and H. Lyn Miles, eds. 1997. *Anthropomorphism, anecdotes and animals.* Albany: SUNY Press.

Mitman, G. 1999. *Reel nature: America's romance with wildlife on screen.* Cambridge, MA: Harvard University Press.

Miyadi, D. 1964. "Social life of Japanese monkeys." *Science* 143:783–86.

Mohnot, S. M. 1971. "Some aspects of social change and infant killing in the Hanuman langur, *Presbytis entellus* (Primates: Cercopithecidae) in western India." *Mammalia* 35:175–98.

———. 1980. "Intergroup infant kidnapping in Hanuman langur." *Folia Primatologica* 34:259–77.

Mohnot, S., M. Gadgil, and S. Makwana. 1981. "On the dynamics of the Hanuman langur populations of Jodhpur (Rajasthan, India)." *Primates* 22:182–91.

Montgomery, Georgina. 2005. "Place, practice and primatology: Clarence Ray Carpenter, primate communication and the development of field methodology." *Journal of the History of Biology* 38:495–533.

Moore, Jim. 1999. "Population density, social pathology and behavioral ecology." *Primates* 40:5–26.

Moos, R., J. Rock, and W. Salzert. 1985. "Infanticide in gelada baboons (*Theropithecus gelada*)." *Primates* 26:497–500.

Mori, Akio, Gurja Belay, and Toshitaka Iwamoto. 2003. "Changes in unit structure and infanticide observed in Arsi geladas." *Primates* 44:217–23.

Morris, R. 1983. *Evolution and human nature.* New York: Putnam Press.

Moss, Cynthia. 1988. *Elephant memories: Thirteen years in the life of an elephant family.* London: Elm Tree Books.

Mulkay, Michael. 1997. *The embryo research debate: Science and the politics of reproduction.* Cambridge: Cambridge University Press.

Myers, G. 1990. *Writing biology: Texts in the social construction of scientific knowledge.* Madison: University of Wisconsin Press.

Napier, J. R., and P. H. Napier. 1970. *Old World monkeys: Evolution, systematics, and behavior.* New York: Academic Press.

Nelkin, Dorothy. 1985. *Selling science: How the press covers science and technology.* New York: Freeman.

Neville, M. K. 1972a. "Social relations within troops of red howler monkeys (*Alouatta seniculus*)." *Folia Primatologica* 18:47–77.

———. 1972b. "The population structure of red howler monkeys (*Alouatta seniculus*) in Trinidad and Venezuela." *Folia Primatologica* 17:56–86.

Newton, Paul. 1986. "Infanticide in an undisturbed forest population of Hanuman langurs, *Presbytis entellus.*" *Animal Behaviour* 34:785–89.

———. 1987. "The social organisation of forest Hanuman langurs (*Presbytis entellus*)." *International Journal of Primatology* 8:199–232.

———. 1988. "The variable social organisation of Hanuman langurs (*Presbytis entellus*), infanticide and the monopolisation of females." *International Journal of Primatology* 9:59–77.

Nickles, Thomas, ed. 2002. *Thomas Kuhn.* Cambridge: Cambridge University Press.

Nishida, T., and K. Kawanaka. 1985. "Within-group cannibalism by adult male chimpanzees." *Primates* 26:274–84.

Nissen, Henry. 1951. "Social behavior in primates." In *Comparative psychology,* ed. C. P. Stone, 423–57. New Jersey: Prentice-Hall.

Nolte, Angela. 1955. "Field observations on the daily routine and social behaviour of common Indian monkeys, with special reference to the bonnet monkey (*Macaca radiata* Geoffroy)." *Journal of the Bombay Natural History Society* 53:177–84.

Onderdonk, Daphne. 2000. "Infanticide of a newborn black-and-white colobus monkey (*Colobus guereza*) in Kibale National Park, Uganda." *Primates* 41:209–12.

Ophir, A., and S. Shapin. 1991. "The place of knowledge: A methodological survey." *Science in Context* 4:3–21.

Oppenheimer, J. R. 1977. "*Presbytis entellus,* the Hanuman langur." In *Primate conservation,* ed. Prince Rainier III of Monaco and Geoffrey Bourne, 469–512. London: Academic Press.

Oppenheimer, J. R., and E. C. Oppenheimer. 1973. "Preliminary observations of *Cebus niger* (Primates: Cebidae) on the Venezuelan coast." *Folia Primatologica* 19:409–36.

Packer, Craig. 1978. "The relative importance of group selection in the evolution of primate societies." *Symposia for the Study of Human Biology* 18:103–12.

———. 1980. "Male care and exploitation of infants in *Papio anubis.*" *Animal Behaviour* 28:512–20.

———. 2000. "Infanticide is no fallacy." *American Anthropologist* 102:829–31.

Packer, Craig, and Anne Pusey. 1983. "Adaptations of female lions to infanticide by incoming males." *American Naturalist* 121:716–28.

Packer, Craig, D. Scheel, and Anne E Pusey. 1990. "Why lions form groups: Food is not enough." *American Naturalist* 136:1–19.

Packer, Craig, et al. 1988. "Reproductive success of lions." In *Reproductive success: Studies of individual variation in contrasting breeding systems.* ed. T. H Clutton-Brock, 363–83. Chicago: University of Chicago Press.

Palombit, R., D. L. Cheney, and R. M. Seyfarth. 1997. "The adaptive role of 'friendship' to female baboons: Experimental and observational evidence." *Animal Behaviour* 54:599–614.

Parmigiani, Stephano, and Frederick vom Saal, eds. 1994. *Infanticide and parental care.* New York: Harwood.

Patterson, I. A. P., et al. 1998. "Evidence for infanticide in bottle nose dolphins: An explanation for violent interactions with harbour porpoises?" *Proceedings of the Royal Society of London*, ser. B, 265:1167–70.

Pereira, M. E., and M. L. Weiss. 1991. "Female mate choice, male migration, and the threat of infanticide in ringtailed lemurs." *Behavioral Ecology and Sociobiology* 28:141–52.

Petter, J. J. 1965. "The lemurs of Madagascar." In *Primate behavior: Field studies of monkeys and apes*, ed. Irven DeVore, 292–319. New York: Holt, Rinehart and Winston.

Peyton, John. 2001. *Solly Zuckerman: A scientist out of the ordinary*. London: John Murray.

Philips, Ruth. 1994. "Why not tourist art? Significant silences in Native American museum representations." In *After colonialism: Imperial histories and postcolonial displacements*, ed. Gyan Prakash, 98–125. Princeton, NJ: Princeton University Press.

Pierotti, R. 1991. "Infanticide versus adoption: An intergenerational conflict." *American Naturalist* 138:1140–58.

Plutchik, R. 1964. "The study of social behaviour in primates." *Folia Primatologica* 2:67–92.

Poirier, Frank. 1969. "The Nilgiri langur (*Presbytis johnii*) troop: Its composition, structure, function and change." *Folia Primatologica* 10:20–47.———. 1972. "The St. Kitt's green monkey (*Cercopithecus aethiops sabaeus*): Ecology, population dynamics and selected behavioral traits." *Folia Primatologica* 17:20–55.

———. 1974. "Colobine aggression: A review." In *Primate aggression, territoriality and xenophobia: A comparative perspective*, ed. Ralph Holloway, 123–57. London: Academic Press.

Poirier, Frank, and E. O. Smith. 1974. "The crab-eating macaques (*Macaca fascicularis*) of Anguar Island, Palau, Micronesia." *Folia Primatologica* 22:258–306.

Porter, Roy. 1978. "Gentlemen and geology: The emergence of a scientific career." *Historical Journal* 21:809–36.

Pusey, Anne E., and Craig Packer. 1987. "The evolution of sex-based dispersal in lions." *Behaviour* 101:275–310.

———. 1994. "Infanticide in lions: Consequences and counter-strategies." In *Infanticide and parental care*, ed. S. Parmigiani and F. S. vom Saal, 277–99. New York: Harwood.

Pusey, Anne E., Jennifer Williams, and Jane Goodall. 1997. "The influence of dominance rank on the reproductive success of female chimpanzees." *Science* 277:828–31.

Radick, Greg. 2005. "Primate language and the playback experiment, in 1890 and 1980." *Journal of the History of Biology* 38:461–93.

———. 2006. "What's in a name? The vervet predator calls and the limits of the Washburnian synthesis." *Studies in the History and Philosophy of the Biological and Biomedical Sciences* 37 (2): 334–62.

Rahaman, H. 1973. "The langurs of the Gir Sanctuary (Gujarat)—a preliminary survey." *Bombay Journal of Natural History* 70:295–314.

Rajpurohit, L. S., and S. M. Mohnot. 1988. "Fate of ousted male resident of one male bisexual troop of Hanuman langurs (*Presbytis entellus*) at Jodhpur, Rajasthan (India)." *Human Evolution* 3:309–18.

Ransom, Timothy. 1981. *Beach troop of the Gombe*. Lewisberg, PA: Bucknell University Press.

Reena, M., and M. Ram. 1991. "Departure of juvenile male *Presbytis entellus* from the natal group." *International Journal of Primatology* 12:39–43.

Rees, Amanda. 2001a. "Anthropomorphism, anthropocentrism and anecdote: Primatologists on primatology." *Science, Technology and Human Values* 26:227–47.

———. 2001b. "Practicing infanticide, observing narrative: Controversial texts in a field science." *Social Studies of Science* 31:507–31.

———. 2006a. "A place that answers questions: Primatological field sites and the making of authentic observations." *Studies in the History and Philosophy of the Biological and Biomedical Sciences* 37 (2): 311–33.

———. 2006b. "Ecology, biology and social life: Explaining the origins of primate sociality." *History of Science* 44:409–34.

———. 2007. "Reflections on the field—primatology, popular science and the politics of personhood." *Social Studies of Science* 37:881–907.

Reynolds, Vernon. 1963. "An outline of the behavior and social organisation of forest living chimpanzees." *Folia Primatologica* 1:95–102.

———. 1965. *Budongo: A forest and its chimpanzees.* London: Methuen.

———. 1967. *Apes: The gorilla, chimpanzee, orangutan and gibbon; their history and their world.* London: Cassell.

———. 1975. "How wild are the Gombe chimpanzees?" *Man* 10:123–25.

Reynolds, Vernon, and Frances Reynolds. 1965. "Chimpanzees of the Budongo forest." *In Primate behavior: Field studies of apes and monkeys,* ed. Irven DeVore, 368–424. New York: Holt, Rinehart and Winston.

Rheingold, H., ed., 1963. *Maternal behavior in mammals.* New York: Wiley.

Richard, A. F., and S. R. Schulman. 1982. "Sociobiology: Primate field studies." *Annual Review of Anthropology* 11:231–55.

Richards, E. 1996. "(Un)boxing the monster." *Social Studies of Science* 26:323–56.

Rijksen, H. 1981. "Infant killing: A possible consequence of a disputed leader role." *Behaviour* 78:138–68.

Ripley, Suzanne. 1967. "Inter-troop encounters between Ceylon gray langurs *(Presbytis entellus)*." In *Social communication among primates,* ed. Stuart A. Altmann, 237–53. Chicago: University of Chicago Press.

———. 1980. "Infanticide in langurs and man: Adaptive advantage or social pathology?" In *Biosocial mechanisms of population regulation,* ed. M. N. Cohen et al., 349–380. New Haven, CT: Yale University Press.

Robbins, David. 1987. "Sport, hegemony and the middle class." *Theory, Culture and Society* 4:579–601.

Roonwal, M., and S. Mohnot. 1977. "Hanuman langur." In *Primates of South Asia: Ecology, sociobiology, and behavior,* ed. M. L. Roonwal and S. M. Mohnot, 234–70. Cambridge, MA: Harvard University Press.

Ross, Andrew, ed. 1996. *Science wars.* Durham, NC: Duke University Press.

Roth, Wolff-Michael, and G. Michael Bowen. 1999. "Digitizing lizards: The topology of 'vision' in ecological fieldwork." *Social Studies of Science* 29:719–64.

Rowell, Thelma E. 1966. "Forest living baboons in Uganda." *Journal of the Zoological Society of London* 149:344–64.

———. 1967. "Variability in the social organization of wild primates." In *Primate ethology,* ed. Desmond Morris, 219–35. London: Weidenfeld and Nicholson.

———. 1974. "The concept of social dominance." *Behavioral Biology* 11:131–54.

———. 2000. "A few peculiar primates." In *Primate encounters: Models of science, gender, and society,* ed. Shirley Strum and Linda Fedigan, 57–70. Chicago: University of Chicago Press.

Rudran, R. 1973. "Adult male replacement in one-male troops of purple-faced langurs (*Presbytis senex senex*) and its effect on population structure." *Folia Primatologica* 19:166–92.

———. 1979. "The demography and social mobility of a red howler (*Alouatta seniculus*) population in Venezuela." In *Vertebrate ecology in the northern Neotropics,* ed. John Eisenberg, 107–26. Washington, DC: Smithsonian Institution Press.

Rudwick, Martin. 1985. *The great Devonian controversy: The shaping of scientific knowledge among gentlemanly specialists.* Chicago: University of Chicago Press.

Sahlins, Marshall. 1959. "The social life of monkeys, apes and primitive man." *Human Biology* 31:54–73.

Schaller, G. B. 1961. "The orang-utan in Sarawak." *Zoologica* 4:73–82.

———. 1965a. "The behavior of the mountain gorilla." In *Primate behavior: Field studies of monkeys and apes,* ed. Irven Devore, 324–67. New York: Holt, Rinehart and Winston.

———. 1965b. "Behavioral comparisons of the apes." In *Primate behavior: Field studies of monkeys and apes,* ed. Irven Devore, 474–81. New York: Holt, Rinehart and Winston.———. 1965c. "Field procedures." In *Primate behavior: Field studies of monkeys and apes,* ed. Irven Devore, 623–29. New York: Holt, Rinehart and Winston.

———. 1965d. *The year of the gorilla.* London: Penguin.

———. 1972. *The Serengeti lion.* Chicago: University of Chicago Press.

Schaller, George B., and John Emlen. 1964. "Observations on the ecology and social behaviour of the mountain gorilla." In *African ecology and human evolution,* ed. F. C. Howell, 368–84. London: Methuen.

Schneirla, T. C. 1950. "The relationship between observation and experimentation in the study of behavior." *Annals of the New York Academy of Sciences* 51:1022–44.

Schubert, G. 1982. "Infanticide by usurper Hanuman langur monkeys: A sociobiological myth." *Social Science Information* 21:199–244.

Schultz, A. H. 1955. "Primatology in its relation to anthropology." *Yearbook of Anthropology* 1955:47–60.

———. 1964. "Primatological symposia of 1962." *Folia Primatologica* 2:119–23.

Scollay, P., and P. DeBold. 1980. "Allomothering in a captive cology of Hanuman langurs (*Presbytis entellus*)." *Ethology and Sociobiology* 1:291–99.

Scott, John P. 1950. "Introduction." *Annals of the New York Academy of Sciences* 51:1003–5.

Scott, Pam, Evelleen Richards, and Brian Martin. 1990. "Captives of controversy: The myth of the neutral social researcher in contemporary scientific controversies." *Science, Technology and Human Values* 15:474–94.

Secord, Anne. 1994. "Corresponding interests: Artisans and gentlemen in nineteenth-century natural history." *British Journal for the History of Science* 27:383–408.

Segerstrale, U. 2000. *Defenders of the truth: The battle for science in the sociobiology debate and beyond.* Oxford: Oxford University Press.

Sekulic, Rana. 1983. "Male relationships and infant deaths in red howler monkeys (*Alouatta seniculus*)." *Zeitschrift für Tierpsychologie* 61:185–202.

Shapin, Steven. 1975. "Phrenological knowledge and the social structure of early 19th century Edinburgh." *Annals of Science* 32:219–43.

———. 1988. "The house of experiment in seventeenth-century England." *Isis* 79:373–404.

———. 1994. *A social history of truth: Civility and science in seventeenth-century England.* Chicago: University of Chicago Press.

Shapin, Stephen, and Simon Schaffer. 1985. *Leviathan and the air-pump: Hobbes, Boyle and the experimental life.* Chicago: University of Chicago Press.

Shea, Christopher. 1999. "Motive for murder?" *Inside Publishing* 9:23–25.

Sherman, Paul. 1981. "Reproductive competition and infanticide in Belding's ground squirrels and other animals." In *Natural selection and social behavior,* ed. Richard D. Alexander and Donald W. Tinkle, 311–31. New York: Chiron Press.

Sibum, Otto. 1995. "Reworking the mechanical value of heat: Instruments of precision and gestures of accuracy in early Victorian England." *Studies in the History and Philosophy of Science.* ser. A, 26:73–106.

Silk, Joan. 2002. "Practice random acts of aggression and senseless acts of intimidation: The logic of status contests in social groups." *Evolutionary Anthropology* 11:221–25.

Silk, Joan, and Craig Stanford. 1999. "Infanticide article disputed." *Anthropology News,* September, 27–28.

Simon, Bart. 1999. "Undead science: Making sense of cold fusion after the (arti)fact." *Social Studies of Science* 29:61–85.

———. 2002. *Undead science: Science studies and the afterlife of cold fusion.* New Brunswick, NJ: Rutgers University Press.

Simonds, P. E. 1962. "The Japan Monkey Centre." *Current Anthropology* 3:303–5.

———. 1965. "The bonnet macaque in South India." In *Primate behavior: Field studies of monkeys and apes,* ed. Irven Devore, 175–96. New York: Holt, Rinehart and Winston.

Smith, James, and Charles McDougal. 1991. "The contribution of variance in lifetime reproduction to effective population size in tigers." *Conservation Biology* 5 (4): 484–90.

Smuts, Barbara. 1985. *Sex and friendship in baboons.* New York: Aldine.

———. 1987a. "Sexual selection and mate choice." In *Primate societies,* ed. Barbara Smuts et al., 385–99. Chicago: University of Chicago Press.

———. 1987b. "Gender, aggression and influence." In *Primate societies,* ed. Barbara Smuts et al., 400–412. Chicago: University of Chicago Press.

Smuts, Barbara, Dorothy Cheney, Richard Seyfarth, Richard Wrangham, and Thomas Struhsaker, eds. 1987. *Primate societies.* Chicago: University of Chicago Press.

Smuts, Barbara, and Robert Smuts. 1993. "Male aggression and sexual coercion of females in nonhuman primates and other mammals: Evidence and theoretical implications." *Advances in the Study of Behavior* 22:1–63.

Sokal, Alan. 1996a. "Transgressing the boundaries: Toward a transformational hermeneutics of quantum gravity." *Social Text* 46/47:217–52.

———. 1996b. "A physicist experiments with cultural studies." *Lingua Franca,* May/June, 62–64.

Sokal, Alan, and Jean Bricmont. 2003/1997. *Intellectual impostures: Postmodern philosophers' abuse of science.* London: Profile Books.

Sommer, Volker. 1987. "Infanticide among free-ranging langurs (*Presbytis entellus*) at Jodhpur (Rajasthan/India): Recent observations and a reconsideration of hypotheses." *Primates* 28:163–97.

———. 1994. "Infanticide among the langurs of Jodhpur: Testing the sexual selection hypothesis with a long-term record." *In Infanticide and parental care,* ed. S. Parmigiani and F. S. vom Saal, 155–98. New York: Harwood.

———. 2000. "The holy wars about infanticide. Which side are you on? And why?" In *Infanticide by males and its implications,* ed. Carel van Schaik and Charles Janson, 9–26. Cambridge: Cambridge University Press.

Sommer, Volker, and S. M. Mohnot. 1985. "New observations on infanticides among Hanuman langurs (*Presbytis entellus*) near Jodhpur, Rajasthan, India." *Behavioral Ecology and Sociobiology* 16:245–48.

Sommer, Volker, and L. S. Rajpurohit. 1989. "Male reproductive success in harem troops of Hanuman langurs (*Presbytis entellus*)." *International Journal of Primatology* 10:293–317.

Southwick, Charles. 1962. "Patterns of intergroup social behavior in primates, with special attention to rhesus and howling monkeys." *Annals of the New York Academy of Sciences* 102:436–54.

———, ed. 1963. *Primate social behavior.* Princeton, NJ: Van Nostrand.

Southwick, Charles, M. A. Beg, and M. R. Siddiqi. 1961a. "A population survey of rhesus monkeys in villages, towns and temples of northern India." *Ecology* 42:538–47.

———. 1961b. "A population survey of rhesus monkeys in northern India: II. Transportation routes and forest areas." *Ecology* 42:698–710.

———. 1965. "Rhesus monkeys in north India." In *Primate behavior: Field studies of monkeys and apes,* ed. Irven Devore, 111–59. New York: Holt, Rinehart and Winston.

Southwick, Charles, and F. C. Cadigan. 1972. "Population structure of Malaysian primates." *Primates* 13:1–18.

Southwick, Charles, M. R. Siddiqi, and M. F. Siddiqi. 1970. "Primate populations and biomedical research." *Science* 170:1051–54.

Sterck, E. H. M. 1997. "Determinants of female dispersal in Thomas's langurs." *American Journal of Primatology* 42:179–98.

———. 1998. "Female dispersal, social organization and infanticide in langurs: Are they linked to human disturbance?" *American Journal of Primatology* 44:235–54.

Stewart, Kelly, and Alexander Harcourt. 1987. "Gorillas: Variation in female relationships." In *Primate societies,* ed. Barbara Smuts et al., 155–64. Chicago: University of Chicago Press.

Stocking, George. 1983. *Observers observed: Essays on ethnographic fieldwork.* Madison: University of Wisconsin Press.

———. 1991. *Colonial situations: Essays on the contextualisation of ethnographic knowledge.* Madison: University of Wisconsin Press.

Struhsaker, Thomas. 1967. "Behavior of vervet monkeys and other cercopithecines." *Science* 156:1197–1203.

———. 1969. "Correlates of ecology and social organisation among African cercopithecines." *Folia Primatologica* 11:80–118.

———. 1977. "Infanticide and social organisation in the redtail monkey (*Cercopithecus ascanius schmidti*) in the Kibale Forest, Uganda." *Zeitschrift für Tierpsychologie* 45:75–84.

Struhsaker, Thomas, and Lisa Leland. 1985. "Infanticide in a patrilineal society of red colobus monkeys." *Zeitschrift für Tierpsychologie* 69:89–132.

———. 1987. "Colobines: Infanticide by adult males." In *Primate societies,* ed. Barbara Smuts et al., 83–97. Chicago: University of Chicago Press.

Strum, Shirley. 2001/1987. *Almost human—a journey into the world of baboons.* London: Elm Tree Books.

Strum, Shirley, and Linda Fedigan, eds. 2000. *Primate encounters: Models of science, gender, and society.* Chicago: University of Chicago Press.

Strum, Shirley, Donald Lindberg, and David Hamburg, eds. 1999. *The new physical anthropology.* Upper Saddle River, NJ: Prentice Hall.

Sugiyama, Yukimaru. 1964. "Group composition, population density and some sociological observations of Hanuman langurs (*Presbytis entellus*)." *Primates* 5:7–38.

———. 1965a. "Behavioural development and social structure in two troops of Hanuman langurs (*Presbytis entellus*)." *Primates* 6:213–47.

———. 1965b. "On the social change of Hanuman langurs." *Primates* 6:381–418.

———. 1966. "An artificial social change in a Hanuman langur troop (*Presbytis entellus*)." *Primates* 7:41–72.

———. 1967. "Social organization of Hanuman langurs." In *Social communication among primates,* ed. Stuart A. Altmann, 221–36. Chicago: University of Chicago Press.

———. 1976. "Characteristics of the ecology of the Himalayan langur." *Journal of Human Evolution* 5:249–77.

———. 1984. "Proximate factors of infanticide among langurs at Dharwar: A reply to Boggess." In *Infanticide: Comparative and evolutionary perspectives,* ed. Glenn Hausfater and Sarah Hrdy, 311–14. New York: Aldine.

Sugiyama, Yukimaru, K. Yoshiba, and M. D. Parthasarathy. 1965. "Home range, mating season, male group and intergroup relations in Hanuman langurs (*Presbytis entellus*)." *Primates* 6:63–106.

Sussman, Robert. 1999. *Primate ecology and social structure: Lorises, lemurs and tarsiers.* Needham Heights, MA: Pearson.

———. 2000. "Piltdown man: The father of American field primatology." In *Primate encounters: Models of science, gender, and society,* ed. Shirley Strum and Linda Fedigan, 85–103. Chicago: University of Chicago Press.

———. 2003. *Primate ecology and social structure: New World monkeys.* Needham Heights, MA: Pearson.

Sussman, Robert, and Audrey Chapman. 2004. *The origins and nature of sociality,* New York: Aldine.

Sussman, Robert, James Cheverud, and Thad Bartlett. 1995. "Infant killing as evolutionary strategy: Reality or myth?" *Evolutionary Anthropology* 3:149–51.

Takahata, Yukio. 1985. "Adult male chimpanzees kill and eat a male newborn infant: Newly observed intragroup infanticide and cannibalism in Mahale National Park, Tanzania." *Folia Primatologica* 44:161–70.

Takasaki, Hiroyuki. 2000. "Traditions of the Kyoto school of field primatology in Japan." In *Primate encounters: Models of science, gender, and society,* ed. Shirley Strum and Linda Fedigan, 151–64. Chicago: University of Chicago Press.

Tappen, N. C. 1960. "Problems of distribution and adaptation of the African monkeys." *Current Anthropology* 1:91–120.

———. 1964. "Primate studies in Sierra Leone." *Current Anthropology* 5:339–40.

Tarara, Erna Burger. 1987. "Infanticide in a chacma baboon group." *Primates* 28:267–70.

Thomas, Marion. 2006. "Yerkes, Hamilton and the experimental study of the ape mind: From evolutionary psychiatry to eugenic politics." *Studies in the History and Philosophy of the Biological and Biomedical Sciences* 37 (2): 273–94.

Thompson, N. S. 1967. "Primate infanticide: A note and a request for information." *Laboratory Primate Newsletter* 6:18–19.

Thorington, R. W. Jr., and C. P. Groves. 1970. "An annotated classification of the Cercopithecoidea." In *Old World monkeys: Evolution, systematics, and behavior,* ed. J. R. Napier and P. H. Napier, 629–47. New York: Academic Press.

Tobias, P. V. 1961. "The work of the gorilla research unit in Uganda." *South African Journal of Science* 57:297–98.

Tomas, D. 1991. "Tools of the trade: Production of ethnographic observations on the Andaman Islands." In *Colonial situations: Essays on the contextualization of ethnographic knowledge,* ed. George Stocking, 75–108. Madison: University of Wisconsin Press.

Traweek, Sharon. 1988. *Beamtimes and lifetimes: The world of high energy physicists.* Cambridge, MA: Harvard University Press.

Treves, A. 1998. "Primate social systems: Conspecific threat and coercion-defence hypotheses." *Folia Primatologica* 69:81–88.

Trivers, Robert. 1972. "Parental investment and sexual selection." In *Sexual selection and the descent of man,* ed. B. Campbell, 136–79. Chicago: Aldine.

———. 1974. "Parent-offspring conflict." *American Zoologist* 14:249–64.

Tucker, Jennifer. 1996. "Voyages of discovery on oceans of air: Scientific observation and the image of science in an age of 'Balloonacy.'" *Osiris* 11:144–76.

Valderrama, X., S. Srikosamatara, and J. G. Robinson. 1990. "Infanticide in wedge-capped capuchin monkeys, *Cebus olivaceus.*" *Folia Primatologica* 54:171–76.

Van Hooff, Jan A. R. A. M. 2000. "Primate ethology and socioecology in the Netherlands." In *Primate encounters: Models of science, gender, and society,* ed. Shirley Strum and Linda Fedigan, 116–37. Chicago: University of Chicago Press.

Van Schaik, Carel. 1996. "Social evolution in primates: The role of ecological factors and male behaviour." *Proceedings of the British Academy* 88:9–31.

———. 2000. "Social counterstrategies against infanticide by males in primates and other mammals." In *Primate males,* ed. P. M. Kappeler, 34–52. Cambridge: Cambridge University Press.

Van Schaik, Carel, P. R. Assink, and N. Salafsy. 1992. "Territorial behavior in Southeast Asian langurs: Resource defense or mate defense?" *American Journal of Primatology* 26:233–42.

Van Schaik, Carel, and R. I. M. Dunbar. 1990. "The evolution of monogamy in large primates: A new hypothesis and some crucial tests." *Behaviour* 115:30–62.

Van Schaik, Carel, and Charles Janson, eds. 2000. *Infanticide by males and its implications.* Cambridge: Cambridge University Press.

Van Schaik, Carel, and P. M. Kappeler. 1997. "Infanticide risk and the evolution of male-female associations in primates." *Proceedings of the Royal Society of London,* ser. B, 264:1687–94.

Veiga, José. 1993. "Prospective infanticide and ovulation retardation in free-living house sparrows." *Animal Behaviour* 45:43–46.

Vogel, Christian. 1971. "Behavioural differences of *Presbytis entellus* in two different habitats." In *Proceedings of the Third International Congress of Primatology,* ed. Hans Kummer, 3:41–47.

———. 1973. "Hanuman as a research object for anthropologists—field studies of social behavior among the gray langurs of India." In *German scholars on India: Contributions to Indian studies,* ed. Cultural Department of the Embassy of the Federal Republic of Germany, 350–65. Varanasi: Chowkhamba Sanscrit Series Office.

———. 1977. "Ecology and sociology of *Presbytis entellus.*" In *Use of nonhuman primates in biomedical research,* ed. M. R. D. Prasad and T. C. Anand, 24–45. New Delhi: Indian National Science Academy.

———. 1988. "Sociobiology of Hanuman langurs (*Presbytis entellus*): Introduction into the Jodhpur field project." *Human Evolution* 3:217–26.

Vogel, Christian, and Hartmut Loch. 1984. "Reproductive parameters, adult male replacements and infanticide among free-ranging langurs (*Presbytis entellus*) at Jodhpur (Rajasthan), India." In *Infanticide: Comparative and evolutionary perspectives,* ed. Glenn Hausfater and Sarah Hrdy, 237–55. New York: Aldine.

Warren, J. M. 1967. "Discussion of social dynamics." In *Social communication among primates,* ed. Stuart Altmann, 255–57. Chicago: University of Chicago Press.

Washburn, Sherwood. 1951. "The new physical anthropology." *Transactions of the New York Academy of Sciences* 13:258–304.

———. 2000/1966. "Conflict in primate society." Reprinted in *The new physical anthropology*, ed. Shirley Strum, Donald Lindberg, and David Hamburg, 270–76. Upper Saddle River, NJ: Prentice Hall.

Washburn, Sherwood, and Irven DeVore. 1961. "The social life of baboons." *Scientific American* 204:62–71.

———. 1962. "Social behaviour of baboons and early man." In *Social life of early man*, ed. Sherwood Washburn, 91–105. London: Methuen.

Washburn, Sherwood, and David Hamburg. 1965. "The study of primate behavior." In *Primate behavior: Field studies of monkeys and apes*, ed. Irven DeVore, 1–13. New York: Holt, Rinehart and Winston.

Washburn, Sherwood, Phyllis Jay, and Jane Lancaster. 1965. "Field studies of Old World monkeys and apes." *Science* 150:1541–47.

Wasser, S. K., ed. 1983. *Social behavior of female vertebrates.* New York: Academic Press.

Wasser, S. K., and D. P. Barash. 1981. "The selfish 'allomother': A comment on Scollay and DeBold (1980)." *Ethology and Sociobiology* 2:91–93.

Watts, David. 1989. "Infanticide in mountain gorillas: New cases and a reconsideration of the evidence." *Ethology* 81:1–18.

———. 1991. "Mountain gorilla reproduction and sexual behavior." *American Journal of Primatology* 24:211–26.

Webster, Bayard. 1982. "Infanticide: Animal behavior scrutinized for clues to humans." *New York Times*, "Science Times," August 17, C1, C4.

Williams, George C. 1966. *Adaptation and natural selection: A critique of some current evolutionary thought.* Princeton, NJ: Princeton University Press.

Wilson, E. O. 1975a. *Sociobiology: The new synthesis.* Cambridge, MA: Harvard University Press.

———. 1975b. "For sociobiology." *New York Review of Books* 22 (20): 60–61.

Winkler, Paul. 1988. "Troop history, female reproductive strategies and timing of male change in Hanuman langurs *(Presbytis entellus)*." *Human Evolution* 3 (4): 227–37.

Winkler, Paul, Hartmut Loch, and Christian Vogel. 1984. "Life history of Hanuman langurs (*Presbytis entellus*): Reproductive parameters, infant mortality and troop development." *Folia Primatologica* 43:1–23.

Wolf, K., and John Fleagle. 1977. "Adult male replacement in a group of silvered leaf monkeys (*Presbytis cristata*) at Kuala Selangor, Malaysia." *Primates* 18:949–55.

Wrangham, Richard. 1980. "An ecological model of female-bonded primate groups." *Behaviour* 75:262–99.

———. 1987. "Evolution of social structure." In *Primate societies*, ed. Barbara Smuts et al., 282–97. Chicago: University of Chicago Press.

———. 1997. "Subtle, secret female chimpanzees." *Science* 277:774–75.

Wright, Robert. 1994. *The moral animal: Evolutionary psychology and everyday life*, New York: Vintage Books.

Würsig, Bernd. 1989. "Cetaceans." *Science* 244:1550–57.

Wynne-Edwards, V. C. 1962. *Animal dispersal in relation to social behaviour.* London: Oliver and Boyd.

———. 1965. "Self-regulating systems in populations of animals." *Science* 147:1543–48.

Yamamura, Norio, et al. 1990. "Why mothers do not resist infanticide: A cost-benefit genetic model." *Evolution* 44 (5): 1346–57.

Yerkes, Robert M. 1943. *Chimpanzees: A laboratory colony.* New Haven, CT: Yale University Press.

———. 1964. Foreword. In *Naturalistic behavior of nonhuman primates,* by Clarence Ray Carpenter. University Park: Pennsylvania State University Press.

Yerkes, Robert M., and Ada W Yerkes. 1929. *The great apes: A study of anthropoid life.* New Haven, CT: Yale University Press.

Yoshiba, K. 1968. "Local and intertroop variability in ecology and social behavior of common Indian langurs." In *Primates: Studies in adaptation and variability,* ed. Phyllis Jay, 217–42. New York: Holt, Rinhart and Winston.

Zuckerman, Solly. 1981/1932. *The social life of monkeys and apes.* London: Routledge and Kegan Paul.

INDEX